NUCLEAR FUEL CYCLE REQUIREMENTS

and supply considerations, through the long-term

February 1978

A report by an expert group
of the
OECD Nuclear Energy Agency

ORGANISATION FOR ECONOMIC CO-OPERATION AND DEVELOPMENT

The Organisation for Economic Co-operation and Development (OECD) was set up under a Convention signed in Paris on 14th December, 1960, which provides that the OECD shall promote policies designed:

— to achieve the highest sustainable economic growth and employment and a rising standard of living in Member countries, while maintaining financial stability, and thus to contribute to the development of the world economy;
— to contribute to sound economic expansion in Member as well as non-member countries in the process of economic development;
— to contribute to the expansion of world trade on a multilateral, non-discriminatory basis in accordance with international obligations.

The Members of OECD are Australia, Austria, Belgium, Canada, Denmark, Finland, France, the Federal Republic of Germany, Greece, Iceland, Ireland, Italy, Japan, Luxembourg, the Netherlands, New Zealand, Norway, Portugal, Spain, Sweden, Switzerland, Turkey, the United Kingdom and the United States.

The OECD Nuclear Energy Agency (NEA) was established on 20th April 1972, replacing OECD's European Nuclear Energy Agency (ENEA) on the adhesion of Japan as a full Member.

NEA now groups all the European Member countries of OECD and Australia, Canada, Japan, and the United States. The Commission of the European Communities takes part in the work of the Agency.

The primary objectives of NEA are to promote co-operation between its Member governments on the safety and regulatory aspects of nuclear development, and on assessing the future role of nuclear energy as a contributor to economic progress.

This is achieved by:

— *encouraging harmonisation of governments' regulatory policies and practices in the nuclear field, with particular reference to the safety of nuclear installations, protection of man against ionising radiation and preservation of the environment, radioactive waste management, and nuclear third party liability and insurance;*
— *keeping under review the technical and economic characteristics of nuclear power growth and of the nuclear fuel cycle, and assessing demand and supply for the different phases of the nuclear fuel cycle and the potential future contribution of nuclear power to overall energy demand;*
— *developing exchanges of scientific and technical information on nuclear energy, particularly through participation in common services;*
— *setting up international research and development programmes and undertakings jointly organised and operated by OECD countries.*

In these and related tasks, NEA works in close collaboration with the International Atomic Energy Agency in Vienna, with which it has concluded a Co-operation Agreement, as well as with other international organisations in the nuclear field.

CONTENTS

APPENDICES

FOREWORD

The OECD Nuclear Energy Agency and the International Atomic Energy Agency have for many years published a joint report entitled, "Uranium Resources, Production and Demand", and a revised edition of this work, dated December 1977, is now available.

This report, on the other hand, is the result of a separate study of the supply and demand outlook for all fuel cycle services, * as well as for uranium, through the long-term. The work was undertaken by the Nuclear Energy Agency's Working Party on Uranium Demand, whose members are listed in Appendix III.

The intent here has been to contribute to the orderly development of nuclear power, by:

1. identifying potential problems in the supply of uranium and fuel cycle services, and possible areas for international co-operation in the resolution of such problems;
2. examining several long-range scenarios to determine the comparative needs of advanced reactors for uranium and for supporting services, thereby establishing the basis for the further development of uranium resources and specific reactor systems; and
3. assisting those having responsibilities in planning, forecasting, and programme management.

This report is the work of a group of technical experts and does not necessarily reflect official policy or endorsement of the report's projections and conclusions by the Member Governments of the Nuclear Energy Agency.

ACKNOWLEDGEMENT

The NEA Working Party on Uranium Demand acknowledges with thanks the calculation by one of its members, Mr. M.J. Crijns of N.V. KEMA, of all demand data presented in this report.

* Except waste management.

I

INTRODUCTION

The nuclear fuel cycle encompasses the procurement and preparation of fuels for use in nuclear power reactors, their recovery and recycle after irradiation, and the safe disposal of all waste generated in these operations.

Over the past two decades, a considerable number of power reactors concepts have evolved. Some remain as concepts, while others have progressed to various stages of demonstration, and still others are in large-scale commercial operation. Although some pattern has now been established in the commercialisation of first-generation reactors, this could easily change with the appearance of an entirely new mix of advanced systems in the future. When uncertainty as to kinds of reactors is compounded with uncertainty as to the numbers of such reactors that may eventually be built, it becomes evident that a study of future supply and demand patterns in the fuel cycle could easily become intractable.

This problem has been somewhat mitigated by organising the report into two principal sections: the short-term (up to the year 2000); and the long-term (from 2000 to 2025). The shorter term covers the period normally considered in current national energy plans, and can therefore reflect both present realities and recent trends in the growth of nuclear power. For the longer-term, and its inherently greater uncertainties, a very different approach has been taken. The primary objective in this section is to explore the impact of widely differing (and hypothetical) reactor strategies upon fuel cycle requirements, rather than to attempt to predict a probable scenario.

In both the short and long-term, attempts have also been made to examine selected aspects of the supply outlook for uranium and fuel cycle services. Current and foreseeable problems of supply are discussed, together with possible solutions within the context of international co-operation. Finally, in light of the projected availability and demand for uranium in the long-term, a future course for reactor systems development is recommended. It is emphasized that all recommendations are made solely on the basis of technical and economic judgments of the Working Party on Uranium Demand. Considerations of a political nature, such as the proliferation of nuclear weapons, are not addressed.

Despite the very speculative nature of this undertaking, studies of this kind are considered essential to the long-range planning needs of the nuclear power industry, the utilities, and to those providing supporting services.

II

SUMMARY

A. GENERAL

A study of nuclear fuel cycle supply and demand must of necessity begin with an estimate of the growth of nuclear power during the period of interest. Although the basis for this estimate will be discussed in detail, it should be mentioned at this stage that nuclear power is expected to make an increasingly significant contribution to the total energy needs of mankind. This is by no means a unique finding, as study after study has confirmed that the major source for projected increases in power production in this century must be nuclear energy, because no other significant option is available to meet the needs of the OECD countries during this period / 1 /. *

Two different estimates have been made of the future growth of world** nuclear power: one, based on present trends** in energy utilization and supply; and another, assuming an accelerated** need for nuclear capacity based on a possible return to higher rates of increase in energy consumption, and new restrictions in the supply of conventional fuels. These estimates are summarized in Table 1:

Table 1. ESTIMATES OF THE GROWTH OF WORLD NUCLEAR POWER

	INSTALLED CAPACITY, GWe			
	1985	1990	1995	2000
"Present Trend" growth	278	504	750	1 000
"Accelerated" growth	368	700	1 220	1 890

In making these estimates, it has been assumed that the public concerns over nuclear energy, and the growing sensitivity by Governments toward the issue of nuclear weapons proliferation will be accommodated without significant further impact on the "present trend". Should this not be the case, lower nuclear capacity estimates would be likely, particularly beyond 1985.

For the longer-term, extrapolation of earlier growth patterns, together with analyses of population trends and energy usage, lead to estimates of 2,150-6,650 GW(e) by the year 2025. These figures represent extensions into the long-term of "present trend" and "accelerated" estimates, respectively.

The breadth of these estimates is characteristic of the extreme uncertainties in long-range nuclear forecasting. However, the intent in the long-term is to identify general relationships between reactor strategies and fuel cycle needs, rather than to quantify. Within this context, these uncertainties can be tolerated.

---

* The figures in brackets concern the references which are given on page 63.

** These terms, as used in this Report, are defined in Part IV.

B. THE SHORT-TERM (1977-2000)

The nuclear power industry's needs for uranium and for fuel cycle services in the short-term are considerable, and indicate the need for strenuous efforts to expand current capacities. Annual requirements for uranium could increase from less than 30,000 t.U. in 1977 to the following levels:

Table 2. WORLD ANNUAL REQUIREMENTS FOR NATURAL URANIUM IN THE SHORT-TERM
(Spent fuels not recycled)

Thousand tonnes uranium

	1985	1990	1995	2000
"Present Trend" growth	71	102	134	178
"Accelerated" growth	88	156	234	338

A comparison of these requirements with a survey of world production capabilities /3/ for the period to 1990 indicates that "present trend" requirements could be met, but that the "accelerated" case is beyond these projected capabilities.

Uranium requirements are shown on a cumulative basis in Table 3:

Table 3. WORLD CUMULATIVE REQUIREMENTS FOR NATURAL URANIUM IN THE SHORT-TERM
(Spent fuels not recycled)

Thousand tonnes uranium

	1985	1990	1995	2000
"Present Trend" growth	423	873	1,477	2,276
"Accelerated" growth	477	1,107	2,127	3,591

These cumulative requirements are within known world uranium resources, presently estimated to total about 4 million tonnes in the cost category less than $ 130/kg U /2/. However, considerable exploration effort will be required before such quantities of uranium can be classified as Reserves, the resource category normally considered for the purpose of production. An atmosphere must be created, both nationally and internationally, that is conducive to the expenditure of this effort, as well as to the attainment of the requisite production levels.

Requirements for fuel cycle services will also increase considerably in the short-term. There has been an adequate supply of enrichment and fuel fabrication services in the past, and as Table 4 indicates, there are plans for future expansion to keep pace with demand.

The situation with respect to reprocessing is considerably more complex, as there are differences of opinion among some member countries as to the necessity for this service in the short-term. World spent fuel arisings, which will amount to about 2,000 tonnes of heavy metal in the current year, will increase substantially, as shown in Table 5.

The accumulation of these arisings will necessitate the construction of storage facilities in those countries that choose to reject the reprocessing option. Countries seeking to recycle light water reactor fuel and/or to stockpile plutonium for fast breeder programs must proceed with the construction of reprocessing and associated waste disposal facilities.

Table 4. WORLD FUTURE DEMAND FOR ENRICHMENT AND FABRICATION SERVICES, AND PLANS FOR SUPPLY

	1985	1990	1995	2000
Enrichment				
(10^3 tonnes separative work/year)				
Demand*	35-46	55-85	77-134	100-189
Supply**	53-65	59-90	...	...
Fabrication				
(10^3 t U/year)				
Demand*	11-15	19-28	28-48	38-74
Supply**	15	...	...	...

* Assuming that spent fuels are not recycled. Demands are shown for conditions of both "present trend" and "accelerated" nuclear power growth.

** Present world capacities: Enrichment, 17.2 tonnes separative work/year;
Fabrication: 6.4 tonnes uranium/year.

Table 5. WORLD SPENT FUEL ARISINGS IN THE SHORT-TERM ("Present trend"/"accelerated" growth)

	1977	1985	1990	1995	2000
Annual, 10^3 tU/Year	2.2	5.9/6.5	11/14	18/26	27/48
Cumulative, 10 tU	*	33/34	76/88	151/191	269/384

* Cumulative figures are taken from 1977 forward. In addition, approximately 4, 000 tonnes of spent fuel were in storage prior to 1977.

Decisions not to proceed with reprocessing, as well as delays in establishing this industry will not only force the adoption of measures for the temporary and long-term storage of spent fuels, but will preclude substantial recycle of recovered plutonium and uranium to light-water reactors before the 1990s. If reprocessing and recycle were to take place on a current basis, cumulative requirements through the year 2000 for natural uranium and for enriching services would be reduced by approximately 15-20%.

C. THE LONG-TERM (2000-2025)

An analysis of alternative reactor strategies makes it clear that continued reliance on thermal converters in this period will result in rapid depletion of known uranium resources and relegate nuclear power to a minor role in satisfying world energy needs. The use of advanced near-breeders (including those which utilize thorium) can do much to reduce the total requirement for uranium, but these requirements will still result in eventual depletion of known reserves. On the other hand, breeder reactors, if demonstrated to be economically feasible, would provide a virtually inexhaustible source of energy supply within foreseeable extensions of known uranium resources. Table 6 summarizes annual uranium requirements for several long-term reactor strategies:

Table 6. WORLD ANNUAL REQUIREMENTS FOR NATURAL URANIUM BY THE YEAR 2025

Thousand tonnes uranium per year

REACTOR STRATEGIES IN THE LONG-TERM	NUCLEAR POWER GROWTH - EXTENSION INTO LONG TERM OF:	
	"PRESENT TREND"	"ACCELERATED"
Light water reactors (LWR), once-through fuelling	350	990
Light water reactors, spent fuel recycle	210	640
Low-enrichment high temperature reactors (HTR)	290	840
Highly-enriched high temperature reactors (HTR)*	240	660
Heavy water reactor (HWR) on Thorium cycle*	150	540
Fast breeder reactor (FBR), 12 year doubling time** ...	95	40
Fast breeder reactor (FBR), 30 year doubling time	95	260

* Additional thorium resources required.

** Corresponds to an advanced FBR with rapid recycle (nine months from reactor discharge to availability as fuel).

As shown in Table 6, the introduction of the fast breeder (particularly one with a short doubling time) can, by the year 2025, reduce annual requirements for uranium far below the corresponding levels for the year 2000. However, coverter and near-breeder reactor strategies continue to require increasingly greater quantities of uranium throughout the long-term. Consequently, by the year 2025, the cumulative uranium requirements of the breeder have already reached a plateau, while the cumulative requirements of other reactor strategies continue to rise (and in fact could reach levels three times as high in the following twenty-five years).

The cumulative uranium requirements by the year 2025 of alternative long-term reactor strategies are shown on Table 7:

Table 7. WORLD CUMULATIVE REQUIREMENTS FOR NATURAL URANIUM BY THE YEAR 2025

Million tonnes uranium

REACTOR STRATEGIES IN THE LONG-TERM	NUCLEAR POWER GROWTH - EXTENSION INTO LONG TERM OF:	
	"PRESENT TREND"	"ACCELERATED"
Light water reactors, once-through fuelling	9.0	20
Light water reactors, spent fuel recycle	5.8	13
Low-enrichment high temperature reactors (HTR)	7.6	17
Highly-enriched high temperature reactors (HTR)* ...	7.3	16
Heavy water reactor (HWR) on Thorium cycle*	6.4	16
Fast breeder reactor (FBR), 12 year doubling time ...	5.8	7.1
Fast breeder reactor (FBR), 30 year doubling time ...	5.8	9.3

* Additional thorium resources required.

At the present time, about 4 million tonnes of uranium has either been identified or surmised to occur in extensions of known districts. Thus, the requirements shown in Table 7 indicate that in the long-term, all reactor strategies require augmentation of what are presently considered as uranium resources, to a varying degree. A concentration of past exploration effort had led to a concentration of low-cost uranium resources in but a few countries, and very large areas of the world remain to be explored. The prognosis for further discovery is therefore excellent, but as in the short-term, a favourable environment is necessary to encourage the expenditure of the necessary effort.

Requirements for fuel cycle services (with the possible exception of enrichment) will also be substantially greater than in the short-term, and will probably entail new technologies as the result of the introduction of advanced reactor concepts. As spent fuel reprocessing and recycle would be a requirement rather than an option under such circumstances, these services assume much greater significance in the long-term.

III

CONCLUSIONS AND RECOMMENDATIONS

URANIUM SUPPLY AND DEMAND

Although presently-known resources of uranium appear adequate to sustain the anticipated growth of nuclear power into the 1990s, two conditions must be fulfilled if uranium supplies are not to be limiting, either in the short or long term:

1. There must be sufficient incentive to provide the basis for the necessary increases in production capacity; and
2. there must be a continuing increase in the tempo of exploration activity to bring greater quantities of known resources and new discoveries into the "Reserve" category, from which production is normally drawn.

The incentives necessary to assure an adequate uranium supply are largely economic in nature, but it is increasingly evident that political forces can have considerable economic impact in this area. For example, political constraints that limit the freedom to develop, produce and export uranium, can contribute to market instability and induce long-term effects by undermining confidence in the future of nuclear power. Because it is so important that there be an adequate supply of uranium for all countries having the common objective of the peaceful use of nuclear power, it is essential that the entire matter be given proper attention on an international level, and in a framework which facilitates equal access to supplies by all countries meeting common peaceful nuclear energy objectives.

FUTURE REACTOR STRATEGIES

Although there is a favourable prognosis for further uranium discovery, the need for nuclear power to serve as a major, long-term source of energy also implies a need for a reactor strategy that is viable by virtue of its efficient utilization of nuclear fuels. This signifies the earliest possible phasing out of today's converters in favour of advanced reactor systems. Advanced converters (including those using thorium) can do much to conserve uranium as compared to present light water reactors, but they, too, lead to depletion of foreseeable resources. Only the fast breeder can essentially place an upper limit upon the amounts of uranium that may eventually be required. This makes the breeder of particular importance to countries that are neither endowed with uranium or conventional fuels, and of broad significance in the event that "inexhaustible" sources of energy do not become available by the turn of the century.

In the view of the Working Party on Uranium Demand, these facts argue strongly not only for the continued development and commercial demonstration of the breeder, but also for the supporting services required to insure its success. Experience must be gained now in the applicable fuel fabrication, reprocessing and recycle techniques to assure the rapid, self-sustaining growth of nuclear power in the long-term. The importance to this objective of a short doubling time makes it essential that progress in these techniques keep pace with the development of the breeder reactor itself.

It must be noted, however, that the rate of development would be moderated by identification of new uranium resources and that investigation of proliferation-resistant fuel cycles could lead to results that would allow a more unified international effort on advanced converter and breeder reactors.

FUEL CYCLE SERVICES

Future demands for fuel cycle services will pose two major challenges to the industry: to provide adequate capacity, at levels several times those of today; and to adapt to new technologies, required by advanced converter and breeder reactor systems. As noted above, it is essential that fuel cycle services keep pace with demands. But in meeting these challenges, the industry must also cope with problems similar to those faced by uranium suppliers - principally, the question of the future role of nuclear power, stemming from uncertainties as to public acceptance, regulatory criteria, and the attitudes of governments. Similarly, safeguards concerns that have led to the restriction of uranium exports, could affect enrichment, fuel fabrication (where plutonium is involved), and reprocessing. As in the case of uranium, it is undesirable for each country to attempt to establish its own supply position when it could be accomplished more efficiently on a regional basis. It is in the interests of all countries to seek means, through international co-operation and controls, for the performance of all fuel cycle services between countries sharing non-proliferation concerns.

IV

THE SHORT-TERM (1977-2000)

A. ESTIMATES OF THE GROWTH OF NUCLEAR POWER

A study of possible future requirements for uranium and for fuel cycle services must necessarily be based on estimates of the growth of world* nuclear power. The uncertainties in making such estimates are considerable. However, the short-term offers the opportunity to reflect recent trends and current energy planning in various parts of the world, and thereby to mitigate the effects of such uncertainties.

In deriving the growth of nuclear power in the short-term (defined as the period from 1977 to 2000), estimates have been under two very different sets of circumstances, resulting in scenarios in which the growth is either "accelerated" or in accord with the "present trend".

The "accelerated" case reflects the goals of ambitious nuclear power programmes, planned in response to the oil embargo, rising costs, and the possible unavailability of conventional fuels. Although delays have overtaken several of these programmes, this could still occur if there were a return to higher rates of increase in energy consumption and new restrictions in the supply of conventional fuels.

The "present trend" estimate takes cognizance of current patterns of energy utilization and supply, as well as present delays in the construction of new reactors, and generally assumes a continuation of this trend. Thus, it reflects the feeling in several countries that the decision whether or not to place major reliance on nuclear power can be postponed for one to two decades, and for the present, must therefore be regarded as the more realistic estimate. In the sections that follow, data will be presented to substantiate the bases for these estimates, and to compare them with other forecasts.

1. Supporting Data

In order to derive the growth of nuclear power in the short-term, estimates were first made of the world's total energy requirements, and of the portion of these needs likely to be fulfilled by electrical power. Further estimation of nuclear power's possible share of the electrical power market then established the nuclear power forecast. These data were developed by the Working Party on Uranium Demand based on responses to questionnaires on plans for the use of nuclear power /3/, consistent with the terms of the "accelerated" and "present trend" scenarios outlined above. They are shown in Appendix I. **

* The term "world" as used in this report, includes:

1. Member countries of the OECD: Australia, Austria, Belgium, Canada, Denmark, Finland, France, Federal Republic of Germany, Greece, Ireland, Italy, Japan, Luxembourg, Netherlands, New Zealand, Norway, Portugal, Spain, Sweden, Switzerland, Turkey, United Kingdom, United States of America;
2. The following non-OECD countries with plans for nuclear power (but excluding the USSR, Eastern Europe, and the People's Republic of China): Latin America-Argentina, Brazil, Chile, Columbia, Cuba, Mexico, Peru, Uruguay, Venezuela; Middle East and Africa - Algeria, Egypt, Iraq, Israel, Kuwait, Niger, Saudi Arabia, South Africa; Asia and the Far East - Bangladesh, Hong Kong, India, Indonesia, Iran, Korea, Malaysia, Pakistan, Philippines, Singapore, Taiwan, Thailand.

** - Total energy requirements: Table I-1
- Electrical share of total energy, system load factor, and nuclear share of electrical capacity: Figure I-1
- Total electric power requirements and capacities, and nuclear power capacities: Table I-2.

The data utilized are in reasonably close agreement with those of other recent forecasts. Table 8 shows a comparison of the world total energy requirements data of this report and of references /4-7/:

Table 8. COMPARISON OF WORLD PRIMARY ENERGY REQUIREMENTS

EJ (10^{18} joules)

	1985	1990	2000
This Report	245-285	289-360	372-559
K.R. Williams /4/	232-274	286-373	364-562
A. Giraud /5/	260	310	410
R.D. Vaughan, et. al. /6/	240	254	308
"WAES" Report /7/	245-264	-	344-445

While there are greater uncertainties in the longer term as to electric power's share of total energy requirements /8/, the data used for this parameter for OECD countries up to the year 2000 in the present study are similar to those reported by the European Community /9/ and the United States /10/. Electrical power's present share of energy requirements (primary energy basis) is about 30% in OECD Member Countries and 20% in the Non-Member countries considered, for a "world" average of 28%. This average is predicted here to be 31-34% by 1985, rising to 33-37% by 1990 and 35-43% by 2000.

The estimate of the extent to which nuclear energy will supplant conventional fuels in the electrical power market, is at the heart of all nuclear power forecasts. The historical growth of nuclear power has been remarkable, and on the basis of the Fischer-Prye market penetration model /11/ could be predicted to account for 50% of all electric capacity in the OECD by the year 1986. However, there is general agreement among most forecasters that although nuclear power will continue to make substantial inroads, it will do so at a much more modest rate. For instance, it has been predicted /12/ that nuclear power's share of total electricity generated in the United States, which was 9.4% in 1976, will rise to 13% in 1980 and 23% in 1985. Another forecast /10/ indicates that nuclear will account for 32-39% of US electrical capacity by the year 2000, while reference /13/ predicts 36% for Western Europe. The use of nuclear power in electrical power generation is currently at a very low level in the Developing Countries. However, according to reference /14/, it is expected to increase rapidly, accounting for at least 3% of installed capacity in 1980, 14% in 1990, and 27% in the year 2000. The overall world figures used in this report, show nuclear power's share of total electrical capacity rising to 16-17% by 1985, 23-24% by 1990, and 32-39% by the year 2000.

2. Statement of Nuclear Power Growth Estimates

The resulting short-term nuclear power estimate is shown on Figure 1 (on a world-wide basis) and on Figure 2 (on a regional basis). Detailed data are listed in Appendix I. *

It can be observed that although the two estimates characterized by the terms "present trend" and "accelerated" are still reasonably close by 1990 (500 GWe and 700 GWe respectively), they diverge rapidly to differ by almost a factor of 2, by the year 2000. This is characteristic of the mounting degree of uncertainty as the term of a forecast is extended. Another noteworthy feature of these estimates is the steady growth of nuclear power in all regions of the OECD and the world, which is plainly indicative of the assumption that countries will increasingly avail themselves of this source of energy to meet their future needs.

* - Nuclear power growth ("present trend" forecast): Table I-3.
- Reactors in operation and under construction: Table I-4. (for comparison with Table I-3).
- Nuclear power growth ("accelerated" forecast): Table I-5.
- Official responses to questionnaires on nuclear power growth: Table I-6 (for comparison with Table I-5).

Figure 1

WORLD NUCLEAR POWER GROWTH
1977-2000

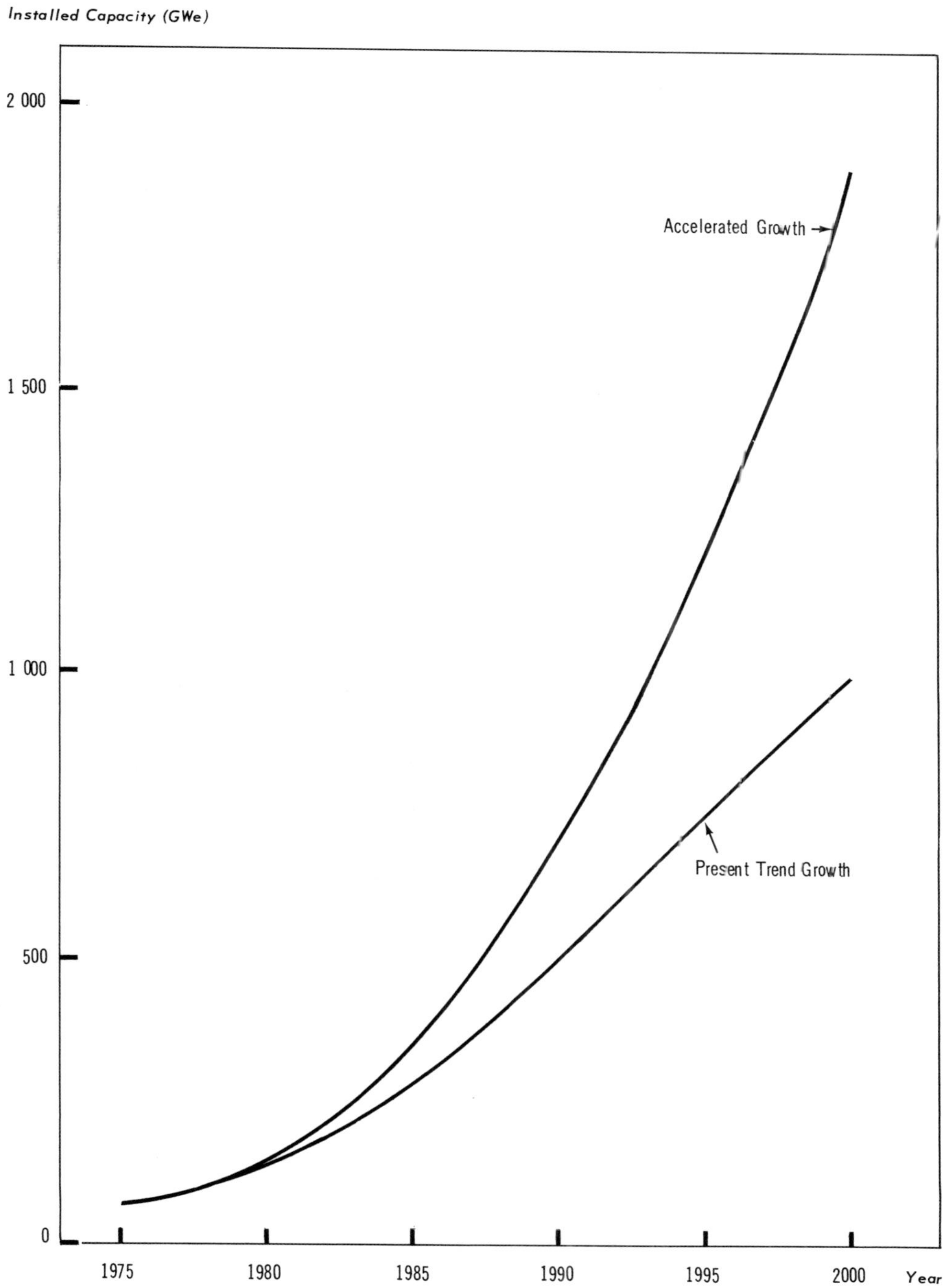

Figure 2

NUCLEAR POWER GROWTH
1977-2000
OECD and non-OECD Countries

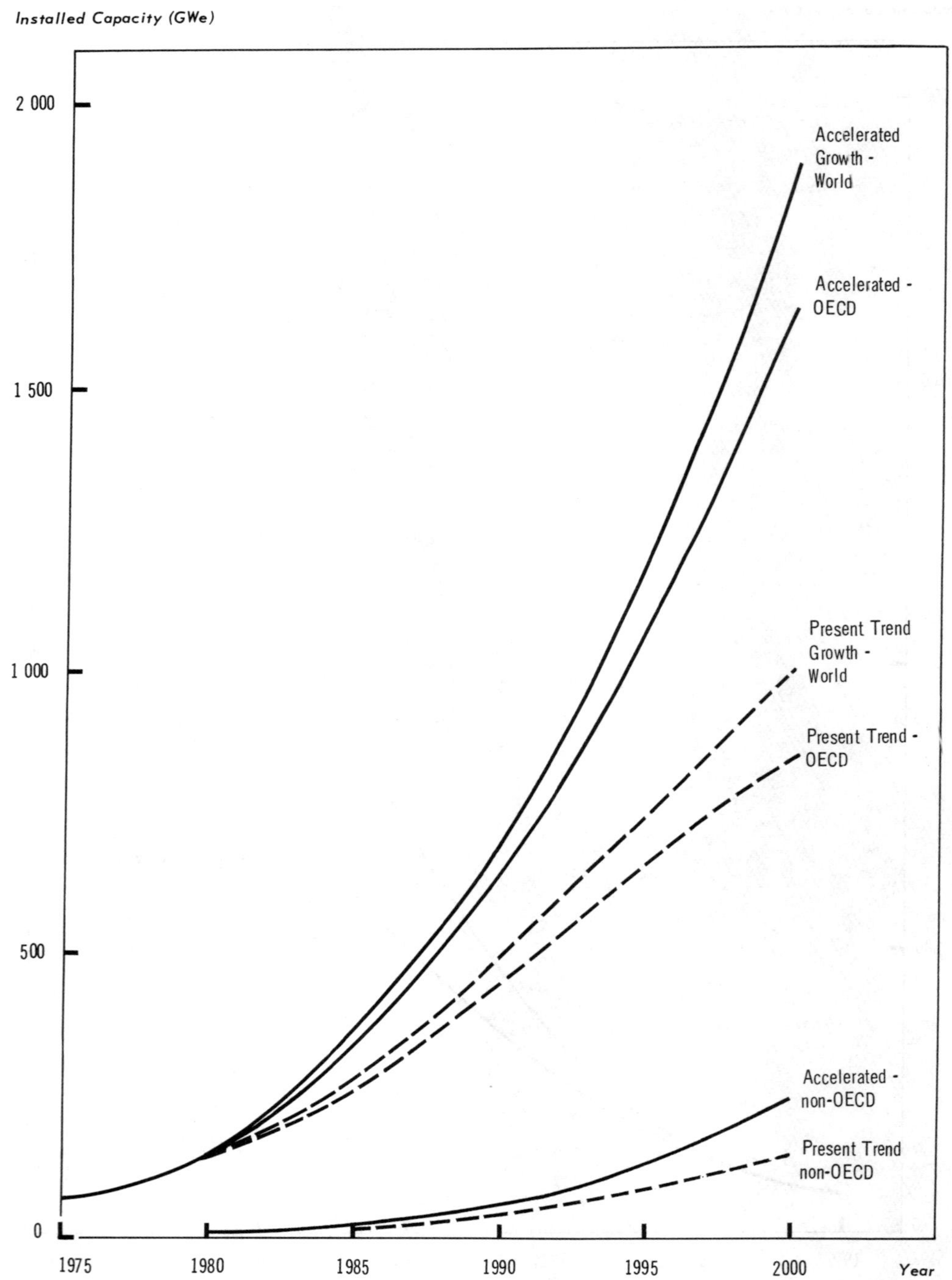

The estimates are also in approximate agreement with others made recently for the purposes of estimating fuel cycle requirements, as shown in Table 9:

Table 9. COMPARISON OF WORLD NUCLEAR POWER GROWTH ESTIMATES

Gigawatts Electrical

	1985	1990	1995	2000
This Report	278-368	504-700	750-1,220	1,000-1,890
K.P. Messer /15/	310-360	550-650	-	-
K.R. Williams /4/	300-350	500-600	730-870	900-1,100
Hanrahan, et. al. /10/	358	620	-	1,410
Braatz and Dibbert /16/	372-456	700-865	1,110-1,360	-
A. Giraud /5/	420	940	1,500	2,100
"WAES" Report /7/	291-412	-	-	913-1,772
M.F. Duret, et. al. /13/	270	-	-	1,141

B. BASIS FOR COMPUTATION OF URANIUM AND FUEL CYCLE REQUIREMENTS

Additional information must be specified to permit computation of world nuclear fuel cycle needs in the short-term, principally, the reactor distribution and reactor fuelling characteristics. These are shown in Appendix I. * The reactor distribution in the short-term assumes continued predominance of light water reactors; a steady increase in the growth of heavy water moderated reactors, accounting for 7.5% of world nuclear capacity by 2000; and the introduction of substantial numbers of fast breeder reactors beginning about 1995. For the "present trend" case, FBR capacity from 1995-2000 roughly follows the historical pattern of LWR introduction from 1964-1969. In the "accelerated" case, FBRs are introduced at a rapid rate, accounting for virtually all new European capacity between 1995-2000. Reactor characteristics have been compiled from data judged to be most representative by the Working Party.

Based on these data, uranium and fuel cycle requirements have been computed by means of the FURE Code /17/, and will be presented in the sections to follow.

C. FUEL CYCLE REQUIREMENTS AND SUPPLY

With the preceeding background, the short-term requirements and supply situation for natural uranium and for fuel cycle services may now be considered.

1. Uranium

a) Demand

Figure 3 shows world short-term requirements for natural uranium on the annual basis. Cumulative requirements are shown on Figure 4, on two bases: either cumulatively up to any given year, or including, in addition, requirements committed for the 30-year lifetime of reactors in-being in any given year. Both figures show the effects of "present trend" or "accelerated" nuclear power growth, and of recycling either uranium, or both uranium and plutonium (beginning in 1985) as compared to no recycling at all. Uranium demand data shown on Figures 3 and 4, are summarized on Table 10.

* Reactor distribution: Table I-7 (world); Table I-8 (OECD countries); Table I-9 (Non-OECD Countries). Reactor characteristics: Table I-10.

Figure 3

WORLD ANNUAL URANIUM REQUIREMENTS
1977-2000

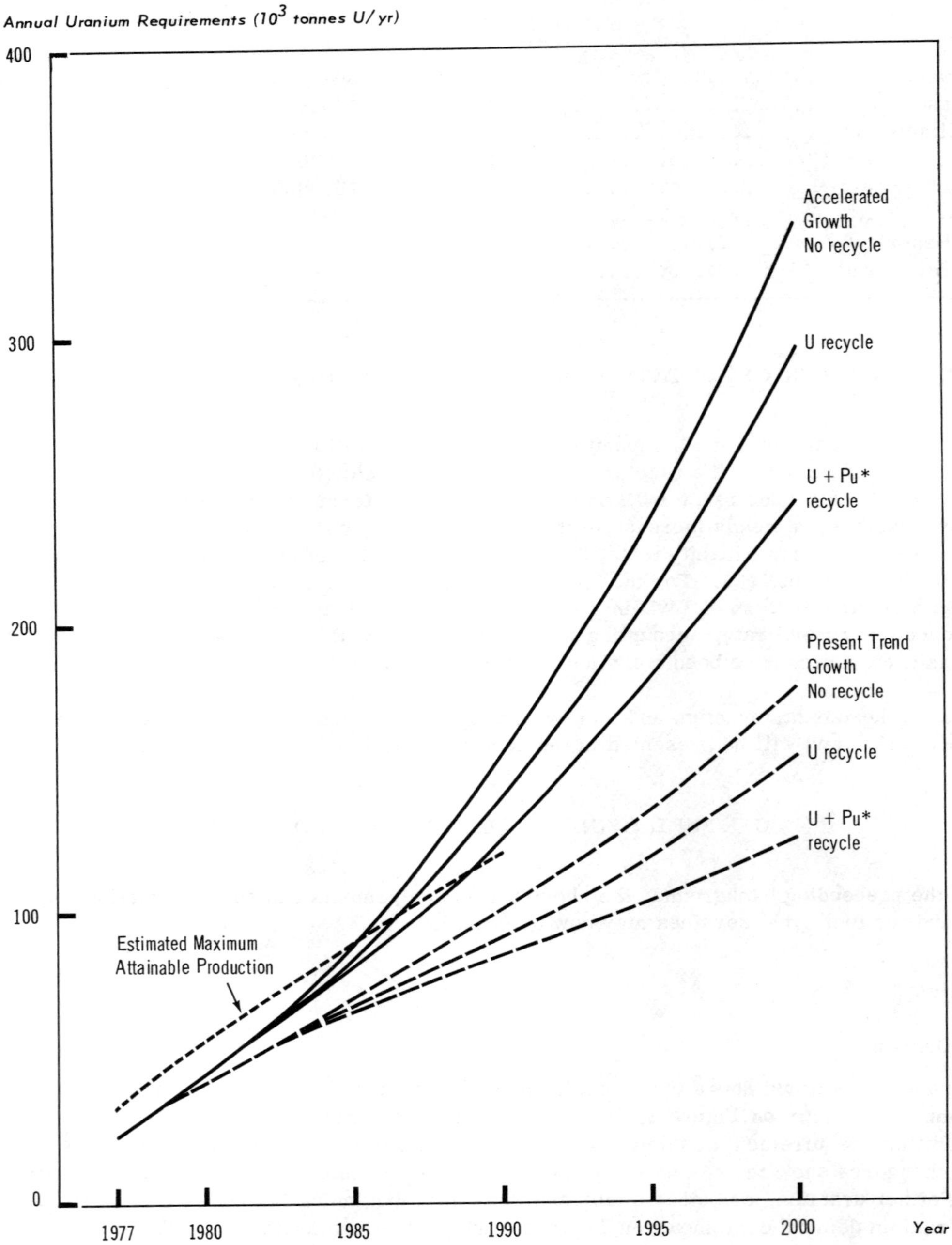

* U-Pu Recycle assumed to begin in 1985.

Figure 4

WORLD CUMULATIVE URANIUM REQUIREMENTS
1977-2000

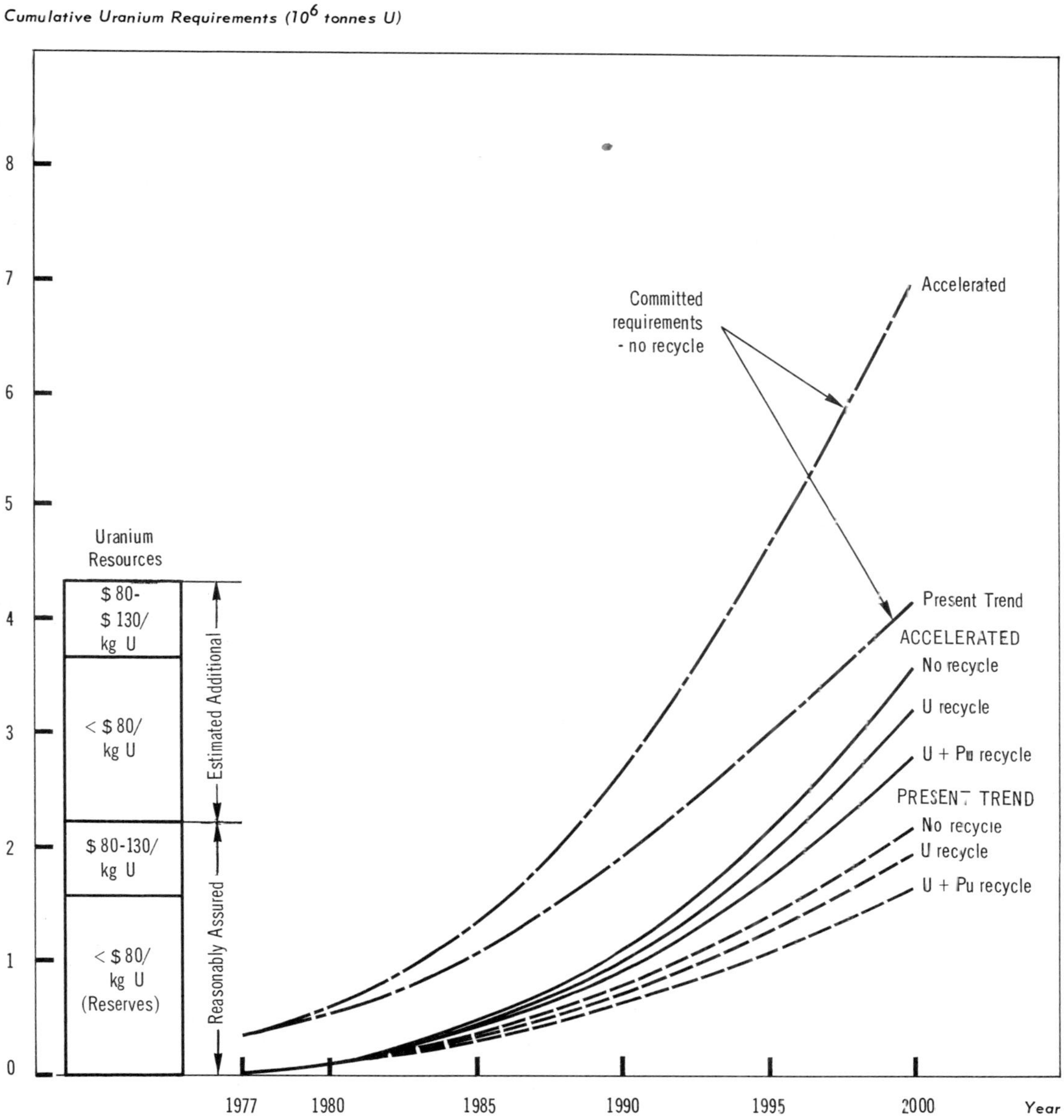

Table 10. WORLD URANIUM REQUIREMENTS, 1977-2000*

1,000 tonnes U

YEAR	WITHOUT RECYCLE				WITH RECYCLE **			
	"ACCELERATED" POWER GROWTH		"PRESENT TREND" POWER GROWTH		"ACCELERATED" POWER GROWTH		"PRESENT TREND" POWER GROWTH	
	ANNUAL	CUMULATIVE	ANNUAL	CUMULATIVE	ANNUAL	CUMULATIVE	ANNUAL	CUMULATIVE
1977	23	23	23	23	23	23	23	23
1978	29	52	29	52	29	52	29	52
1979	35	87	35	87	35	87	35	87
1980	43	130	41	128	43	130	41	128
1981	51	181	47	175	51	181	47	175
1982	60	241	53	228	60	241	53	228
1983	69	310	59	287	67	308	57	285
1984	79	389	65	352	74	382	61	346
1985	88	477	71	423	82	464	65	411
1986	98	575	78	501	90	554	69	480
1987	111	686	84	585	98	652	73	553
1988	125	811	90	675	106	758	77	630
1989	140	951	96	771	115	873	81	711
1990	156	1,107	102	873	126	999	85	796
1991	171	1,278	108	981	137	1,136	89	885
1992	188	1,466	114	1,095	149	1,285	93	978
1993	206	1,672	121	1,216	160	1,445	96	1,074
1994	221	1,893	127	1,343	171	1,616	100	1,174
1995	234	2,127	134	1,477	181	1,797	104	1,278
1996	247	2,374	142	1,619	193	1,990	108	1,386
1997	267	2,641	150	1,769	205	2,195	112	1,498
1998	295	2,936	160	1,929	218	2,413	116	1,614
1999	317	3,253	169	2,098	230	2,643	120	1,734
2000	338	3,591	178	2,276	242	2,885	125	1,859

* Based on 0.25% U-235 enrichment plant tails assay and other data shown in Appendix I.

** Beginning in 1985.

The separate requirements of OECD and non-OECD countries are given on the annual and cumulative bases in Figures 5 and 6, respectively.

Figures 3-6 indicate a wide range of possible demands for natural uranium. Considerable differences can be noted, even within the "present trend" scenario, depending upon the assumptions of spent fuel recycle. Practically speaking, however, it is not considered likely that appreciable amounts of spent fuels will be reprocessed before the mid-1990s. Therefore, the plutonium recovered from these operations is likely to be reserved for use in the initial cores of fast breeder reactors, rather than recycled in light water reactors. Hence, it is probable that uranium requirements will follow the non-recycle curves of Figures 3-6, while the recycle curves indicate (academically) the savings that might have been realized, had reprocessing and recycle taken place on a world-wide basis, as originally planned, starting in the mid-1980s.

Before proceeding to a discussion of the adequacy of uranium supply, two important points must be noted in connection with the demand data presented here. First, all demands were computed on the basis of an 0.25% U-235 enrichment plant tails assay. A variety of contractual schemes have been in effect in the past, including the use of a fictitious tails assay for the purpose of computing natural

Figure 5

ANNUAL URANIUM REQUIREMENTS
1977-2000
OECD and Non-OECD Countries

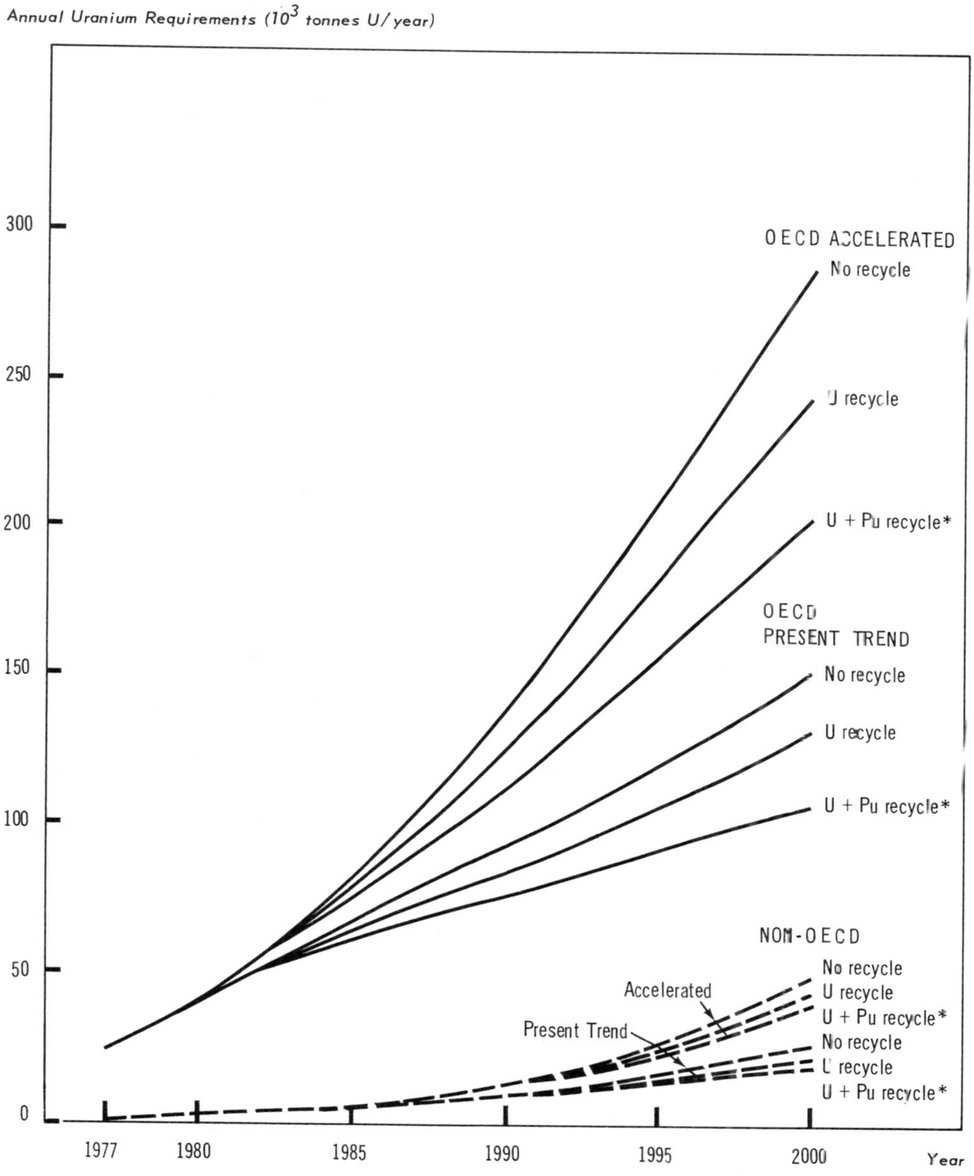

* U + Pu Recycle begins 1985.

Figure 6

CUMULATIVE URANIUM REQUIREMENTS
1977-2000
OECD and Non-OECD Countries

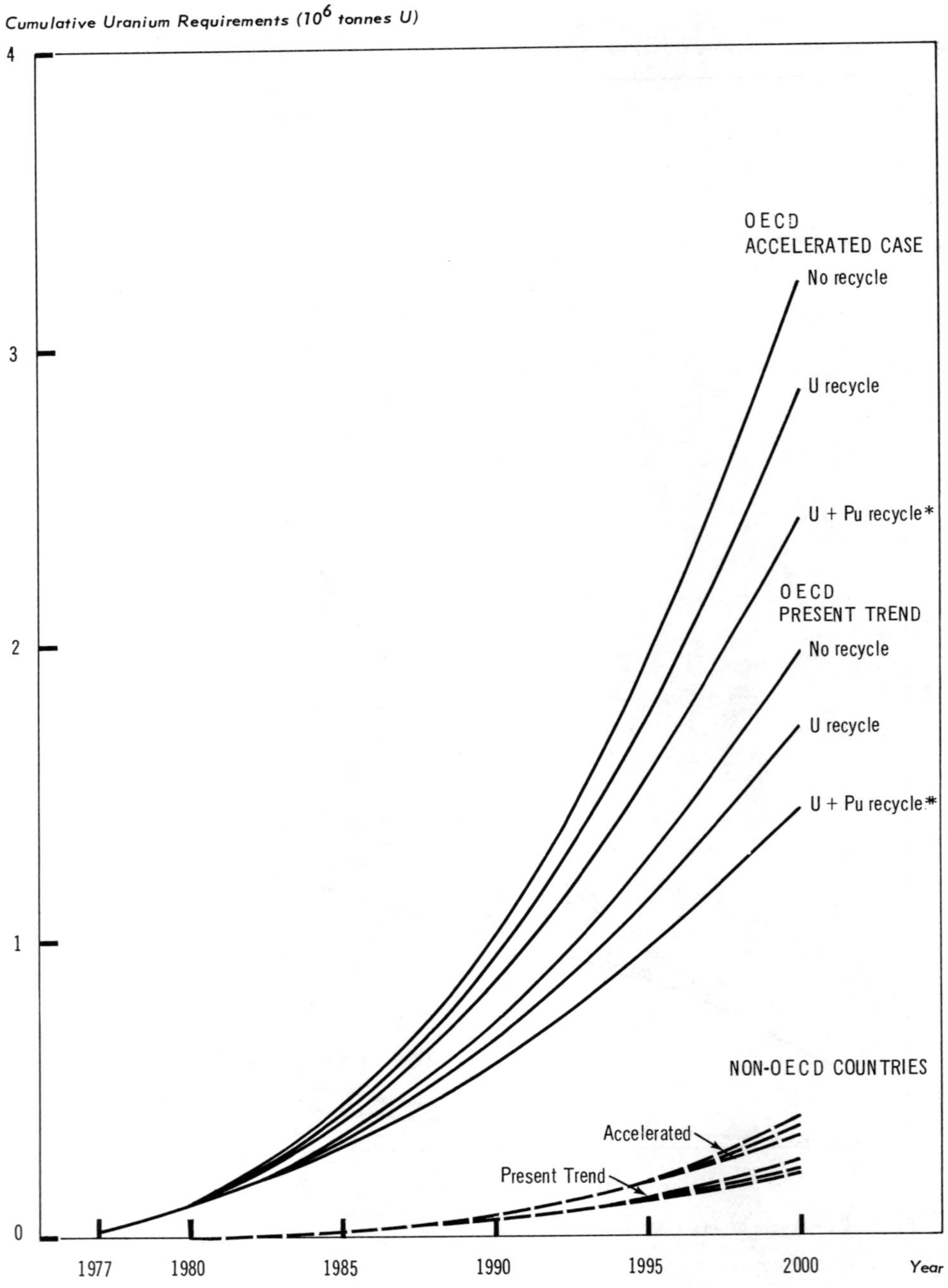

* U + Pu recycle begins 1985.

uranium required of utilities. To the extent that the tails assay will vary in the future, so too will uranium requirements. Although there have been warnings of an upward trend in recent years, present indications are that the tails assay will go no higher than 0.25% U-235. It is also possible that the tails assay could drop if a large amount of enriching capacity were to become available, or with the advent of a newer technology such as laser isotope separation. As an extreme example, a tails assay of 0.1% U-235 would result in a reduction in annual uranium requirements of about 30%. The second point to be noted, is that uranium requirements have been computed here on an "as-needed" basis, with due allowance for the lead time required for mining and milling, chemical conversion, and enriching operations. However, some enrichment contract provisions have in the past required firm commitments for the receipt of enrichment services by utilities, regardless of subsequent changes in reactor schedules. To the extent that this may continue to occur in the future, uranium requirements may be in excess of those stated here.

b) Supply

The latest available world uranium resources figures /2/ show about 2.2 million tonnes of uranium in the "Reasonably Assured" category, and an additional 2.1 million tonnes as "Estimated Additional Resources", each recoverable at costs less than \$ 130/kg U (\$ 50/lb U_3O_8). The distinctions drawn between the two categories of resources, based on differing degrees of geological evidence, make it essential that each be regarded as a discrete entity. However, the resource level of Estimated Additional Resources should be considered as having the potential for later conversion to Reasonably Assured Resources, as the possible result of further exploration effort. Finally, it should be noted that uranium production is normally drawn from "Reserves", now defined as the low-cost (< \$ 80/kg U) subcategory of Reasonably Assured Resources, and estimated at 1.65 million tonnes uranium. As foreseeable demand must be satisfied by Reserves, a prime objective of exploration must be to increase the known quantities of this resource.

Table 11 summarizes world uranium resources in both the "Reasonably Assured" and "Estimated Additional" categories /2/.

Table 11. WORLD URANIUM RESOURCE POSITION*

1,000 tonnes U

	REASONABLY ASSURED RESOURCES		ESTIMATED ADDITIONAL RESOURCES	
	< \$ 80/kg U** (<\$ 30/lb U_3O_8)	\$ 80-130/kg U (\$ 30-50/lb U_3O_8)	< \$ 80/kg U (<\$ 30/lb U_3O_8)	\$ 80-130/kg U (\$ 30-50/lb U_3O_8)
Australia	289	7	44	5
Canada	167	15	392	264
South Africa	306	42	34	38
United States	523	120	838	215
Other African countries ...	218	6	118	10
Western Europe	57	325	41	37
Countries not included above	87	29	44	16
Total (rounded)	1,650	540	1,510	590

* As of January 1, 1977

** Defined as "Reserves".

For purposes of comparing uranium supply and demand, histograms indicating the quantities of uranium resources presently estimated in each of the respective categories are shown on Figure 4 (world cumulative demand). This comparison indicates that Reserves are sufficient to cover actual needs until 1992-1996 (for "accelerated" and "present trend" nuclear power growth, respectively, assuming that spent fuels are not recycled). However, current Reserves may only be sufficient to cover the lifetime requirements of reactors in-being in 1986-1988, under the same set of assumptions. If spent fuels were recycled at the earliest opportunity (about 1985), the stated coverages of demand by Reserves could be extended by another 2 to 3 years. The total of Reasonably Assured Resources could cover demand until 1995-2000, under the two conditions of nuclear power growth, assuming no recycle.

This then raises the question as to whether annual production levels of uranium will be adequate to meet future demands. Table 12 shows estimates of maximum possible rates of production that can be expected from known resources, given a favourable economic climate and proper planning /2/.

Table 12. ATTAINABLE PRODUCTION CAPABILITIES

Tonnes U

	1977	1980	1985	1990
Australia	400	500	11,800	20,000
Canada	6,100	7,950	12,500	11,250
South Africa	6,700	11,700	12,500	12,000
United States	14,700	22,600	36,000	47,000
Western Europe	2,600	3,700	5,600	5,900
Other African Countries ...	2,400	5,300	11,200	11,200
Countries not included above .	400	1,300	2,000	2,300
Total (rounded)	33,000	53,000	92,000	110,000

The total attainable production levels listed on Table 12, have been plotted on Figure 3 for purposes of comparison with world annual uranium requirements in the short-term. It may be observed that projected capabilities for production, which are based on known and anticipated additions to Reserves, could sustain the growth of nuclear power until at least 1990 (in all instances except "accelerated" growth, without recycle). The term "could" is stressed, because as stated in reference /2/, several conditions must be fulfilled before the attainable production capabilities shown on Table 12 can be realized. The necessary conditions, all contributing to the provision of sufficient economic incentive to uranium producers, include:

1. the orderly growth of nuclear power to assure sufficient lead time for planning, and to attract the necessary capital;
2. the absence of market perturbations (e.g., unloading of stocks, changes in enrichment contract terms, etc.) that can distrub the equilibrium between supply and demand; and
3. a reduction in the recent inflationary pressures on plant construction and operation, which have been particularly damaging to the uranium mining and milling industry.

From 1990 to 2000, as shown on Figure 3, there will be increasingly greater demands on annual output. It must be emphasized that one cannot predict now with assurance that supply can keep pace with projected demands for that period, even in a favourable economic climate. One thing that is certain, is that production rates, based on presently-known reserves, must eventually decrease as higher grade ores become depleted and each deposit is mined out. This then places added emphasis upon the necessity for a continually increasing tempo of exploration activity, to bring other resources into the reserve category, and for new discoveries.

Detailed discussions of constraining factors, and of the necessary ingredients for insuring adequate uranium supplies in the nearer term, are contained in references /2/ and /18/.

In the preceding discussion it has been implied that the incentives necessary to insure an adequate uranium supply are economic in nature. In this connection, it should also be noted that political factors may also have considerable impact. On a national scale, these could include the economic effects of regulation of mining activities, and the suspension of resource development while inquiries are conducted on environmental and other concerns, pending the formulation of governmental policies. In addition, there are political factors of an international nature, that are even more significant. For example, supplier countries which together account for almost 70% of the world's known uranium resources, are currently seeking new conditions on the export of uranium satisfactory to their respective safeguards concerns. The immediate effect of a major change in uranium supply policy is to contribute to uranium market instability, as each country or region attempts to establish its own supply position, and over-all confidence in nuclear power is undermined. It is therefore essential that an accommodation of these safeguards concerns be quickly negotiated, preferably through an international control system, before these perturbations unduly constrain the availability of uranium.

2. Enrichment

a) Demand

The short-term annual world demand for enrichment services is shown on Figure 7 and summarized on Table 13. Demands are also show separately for OECD and non-OECD countries on Figure 8. Again,

Table 13. WORLD URANIUM ENRICHMENT REQUIREMENTS, 1977-2000*

1,000 tonnes U

YEAR	WITHOUT RECYCLE				WITH RECYCLE **			
	"ACCELERATED" POWER GROWTH		"PRESENT TREND" POWER GROWTH		"ACCELERATED" POWER GROWTH		"PRESENT TREND" POWER GROWTH	
	ANNUAL	Cumulative	ANNUAL	Cumulative	ANNUAL	Cumulative	ANNUAL	Cumulative
1977	12	12	12	12	12	12	12	12
1978	14	26	14	26	14	26	14	26
1979	16	42	16	42	16	42	16	42
1980	19	61	19	61	19	61	19	61
1981	23	84	22	83	23	84	22	83
1982	28	112	25	108	28	112	25	108
1983	34	146	28	136	33	145	27	135
1984	40	186	31	167	38	183	29	164
1985	46	232	35	202	43	226	31	195
1986	53	285	39	241	49	275	34	229
1987	60	345	43	284	55	330	37	266
1988	68	413	47	331	61	391	40	306
1989	76	489	51	382	68	459	43	349
1990	85	574	55	437	74	533	46	395
1991	94	668	59	496	82	615	49	444
1992	103	771	63	559	88	703	52	496
1993	114	885	68	627	96	799	55	551
1994	124	1,009	72	699	103	902	58	609
1995	134	1,143	77	776	111	1,013	62	671
1996	144	1,287	82	858	118	1,131	65	736
1997	155	1,442	86	944	126	1,257	69	805
1998	166	1,608	91	1,035	134	1,391	72	877
1999	177	1,785	95	1,130	141	1,532	75	952
2000	189	1,974	100	1,230	150	1,682	79	1,031

* Based on 0.25% U-235 enrichment plant tails assay and other data presented in Appendix I.

** Beginning in 1985.

Figure 7

ANNUAL ENRICHMENT REQUIREMENTS OF THE WORLD
1977-2000

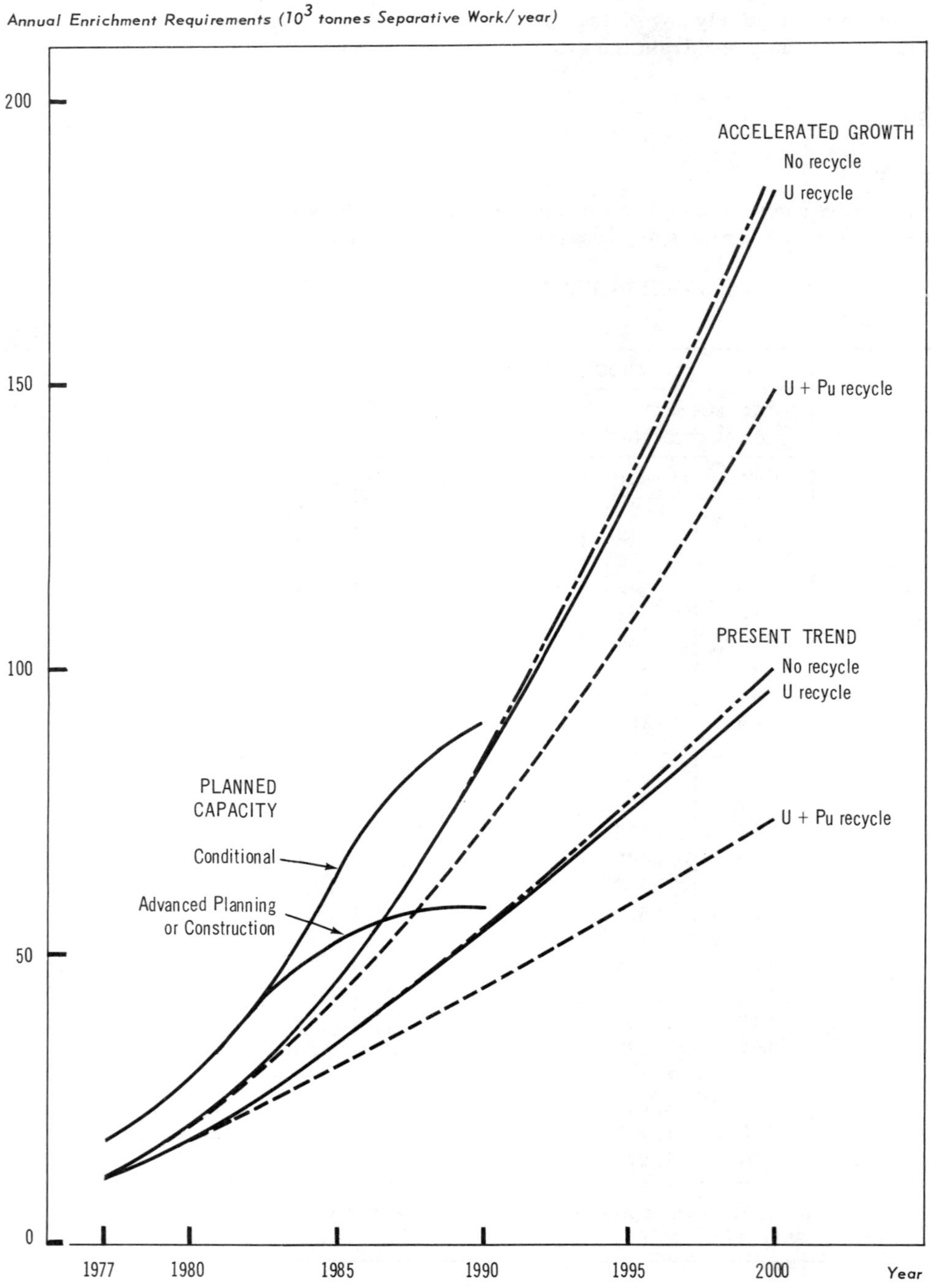

Figure 8

ANNUAL ENRICHMENT REQUIREMENT
1977-2000
OECD and Non-OECD Countries
(no recycle)

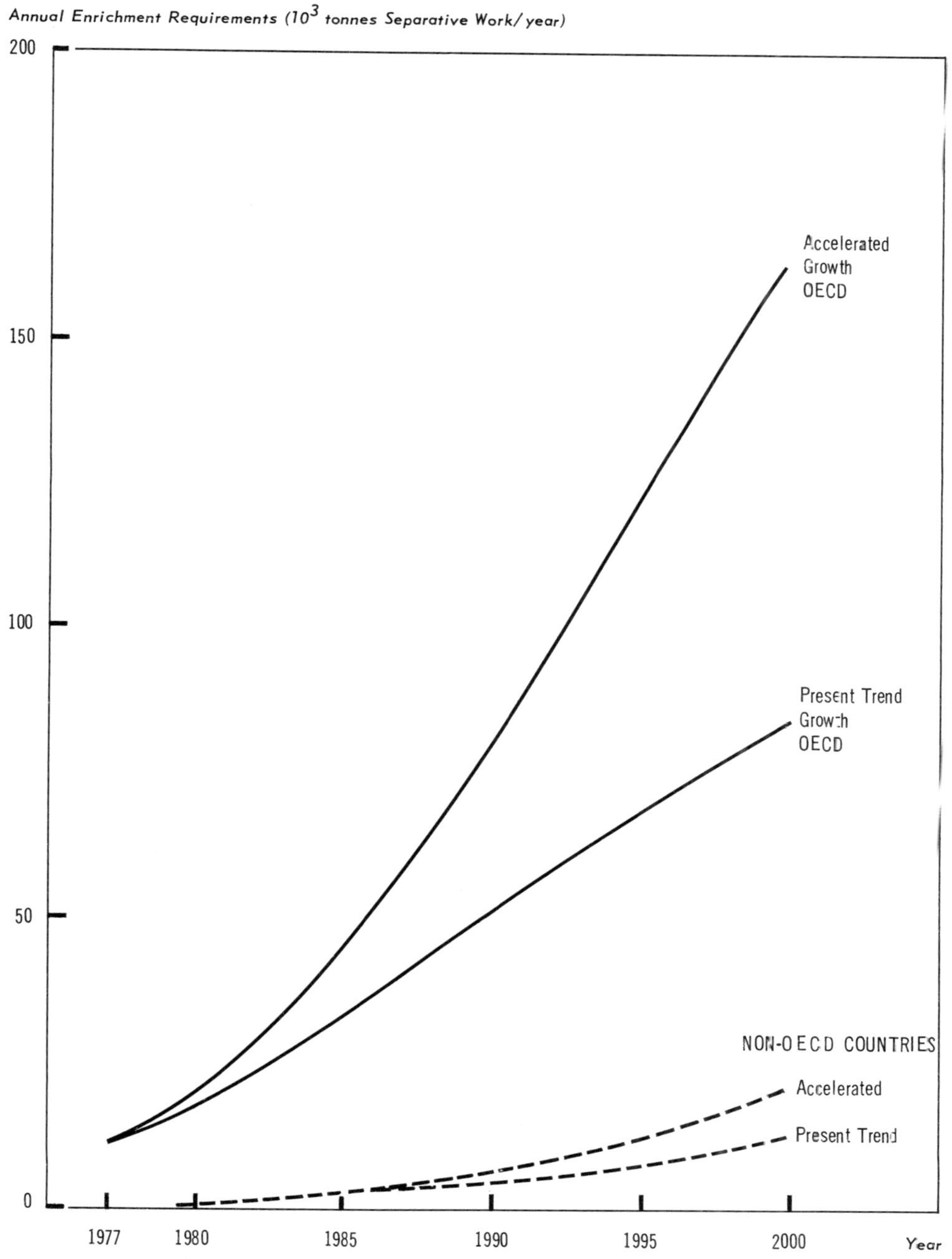

requirements that are computed on the assumption that spent fuels will not be recycled to light water reactors, must be regarded as the most realistic. In this event, the demand for separative workd would be as summarized in Table 4.

It may be noted, as with uranium, that requirements vary considerably between the two scenarios. In addition, the inability to recycle fuel has, as indicated by Figures 7-8, increased the level of requirements for enrichment services.

b) Supply

The current plans for the expansion of the uranium enriching capacities of several countries, as discussed in references /16/ and /19/-/22/, are summarized in Table 9. The statement of a single figure for a given supplier, indicates a relatively firm plan or commitment. Where a range of capacities is stated, the first figure refers to firm plans, while the last figure represents plans for further expansion, if markets should warrant such expansion. However, when the first figure in a given range is zero, the entire project must be regarded as extremely conditional.

Table 14. EXISTING AND PLANNED COMMERCIAL ENRICHMENT CAPACITY IN THE WORLD

10^3 tonnes Separative Work per Year

	1977	1980	1985	1990
United States DOE (Gaseous Diffusion)	17	21	26.6	26.6
United States DOE (Centrifuge)	-	-	8.8	8.8*
Eurodif	-	6.1	10.8	10.8
Coredif	-	-	4.6-5.4	5.4-10.8
Urenco	0.2	0.6	2-5	2-10
South Africa	-	-	0.01-5	5
Brazil	-	-	0.2-2.5	0.2-2.5
Canada	-	-	-	0-9
Australia	-	-	0-0.3	0-0.3
Japan	-	-	0.1	0.1-6
Total	17.2	27.7	53-65	59-90
USSR (export)	1	2	1.5-3	-

* Does not include private enrichment ventures that might eventually materialize.

The planned enrichment capacities shown on Table 14 have been plotted on Figure 7 to provide a comparison of projected supply and demand. This comparison indicates that sufficient capacity is now planned to cover demand at least through 1990, in the event of either "present trend" or "accelerated" nuclear growth and assuming no recycle, with considerable excess capacity before that date. Before passing such a judgment, however, it should be recalled that the demand for enrichment services has been computed here on an "as-needed" basis, taking into account due delays for chemical conversion and fuel fabrication. Although this is the general principle on which United States ERDA (now DOE) "requirements" contracts were once based, the bulk of enrichment contracts are now of the "fixed commitment" type, which call for deliveries on specific schedules, regardless of subsequent delays in reactor construction and operation. In the event that enrichment contracting continues to be of this nature, the actual demand could be higher than shown here, and the adequacy of planned capacity could be diminished accordingly.

However, it is possible to assume an optimistic outlook with respect to the adequacy of supply of uranium enrichment services. Three very positive factors that engender such confidence are the general recognition of the essentiality of these services, the willingness of several goverments to provide them, and the trouble-free operation of plants already in existence. However, as pointed out in reference /23/, there are also factors that could constrain the availability of such services. These include not only technical and economic factors (high specific capital investment and low rates of capital recovery, a large scale factor, market uncertainties and classified technologies), but also the probability that increasingly stringent security conditions and arrangements will apply to the sale of enrichment services as well as end use of the enriched uranium and products derived therefrom. It is therefore concluded /23/ that the task to develop and assure adequate enrichment capacity calls for much greater co-operation, both in the areas of planning and risk-sharing, between countries in possession of enrichment technology, consumer nations, and international organisations.

3. Reprocessing Services

There are considerable differences of opinion in the world today on the need for reprocessing spent fuels. Whether a given country will reprocess spent fuels will likely depend on its approach to the safeguarding of nuclear materials, the adequacy of its supply of natural uranium and other energy resources, and its plans for fast breeder development. As stated earlier, only a modest amount of fuel is expected to be recycled to light water reactors in the short-term. However, although the United States has embarked on a programme for long-term storage of spent fuels pending resolution of its deep non-proliferation concerns, reprocessing remains the avowed objective of several other countries with major nuclear programmes, largely to provide plutonium for fast breeder reactors scheduled for introduction in the 1990s.

a) Spent Fuel Arisings

Spent fuel arisings, or potential reprocessing requirements, are shown on Figures 9 and 10 on the annual and cumulative bases, respectively and summarized on Table 15. It may be observed that the amount of spent fuel currently being discharged into reactor cooling ponds is rapidly increasing, presenting both a storage problem for reactor operators and a potential market for the emerging reprocessing industry. World spent fuel arisings, which will amount to about 2,000 tonnes of heavy metal content in the current year, will increase to about 6,000 tonnes in 1985, 11-14,000 tonnes in 1990, and 27-48,000 tonnes by the year 2000. The cumulative figures are even more impressive, as shown on Tables 5 and 15. These data demonstrate the urgent need to resolve the mounting spent fuel storage problem of reactor operators. In the United States (and possibly in other countries that choose to forego the reprocessing option), industry is urging that the government rapidly develop new regulatory criteria for handling and storing spent fuels in addition to implementing plans for long-time storage in national repositories /24/. Although several other countries having major nuclear programmes intend to establish away-from-reactor storage facilities, these are usually planned in conjunction with reprocessing plants.

b) Availability of Reprocessing Services

A study was made in 1976 of possible future reprocessing capabilities in the OECD countries /25/. Since the publication of the study, the United States has deferred the reprocessing option; hence, its contribution in the short-term to the reprocessing capabilities of the OECD should now be excluded from further consideration, as well as the spent fuel arisings of United States reactors from the total OECD demand for reprocessing services. Thus, it is probably more valid to study possible supply and demand relationships in reprocessing on a regional basis. In Table 16, the European Region is taken as an example.

As shown by Table 7 and noted elsewhere /27/, there is an imbalance between supply and demand for reprocessing services in the European Region. There is even greater imbalance in the OECD Pacific Region, and the needs of the Developing Countries have not even been considered. To illustrate the magnitude of the short-fall on a world-wide basis, the total planned reprocessing capacity is plotted on the demand curves, Figures 9 and 10. Thus, it is estimated that the current (end of 1977) 4,000-tonne world inventory of unreprocessed spent fuels will increase to about 60,000 tonnes by 1990 (based on "present trend" nuclear power growth, and the lower estimate of planned reprocessing capacity).

Figure 9

ANNUAL REPROCESSING REQUIREMENT
1977-2000

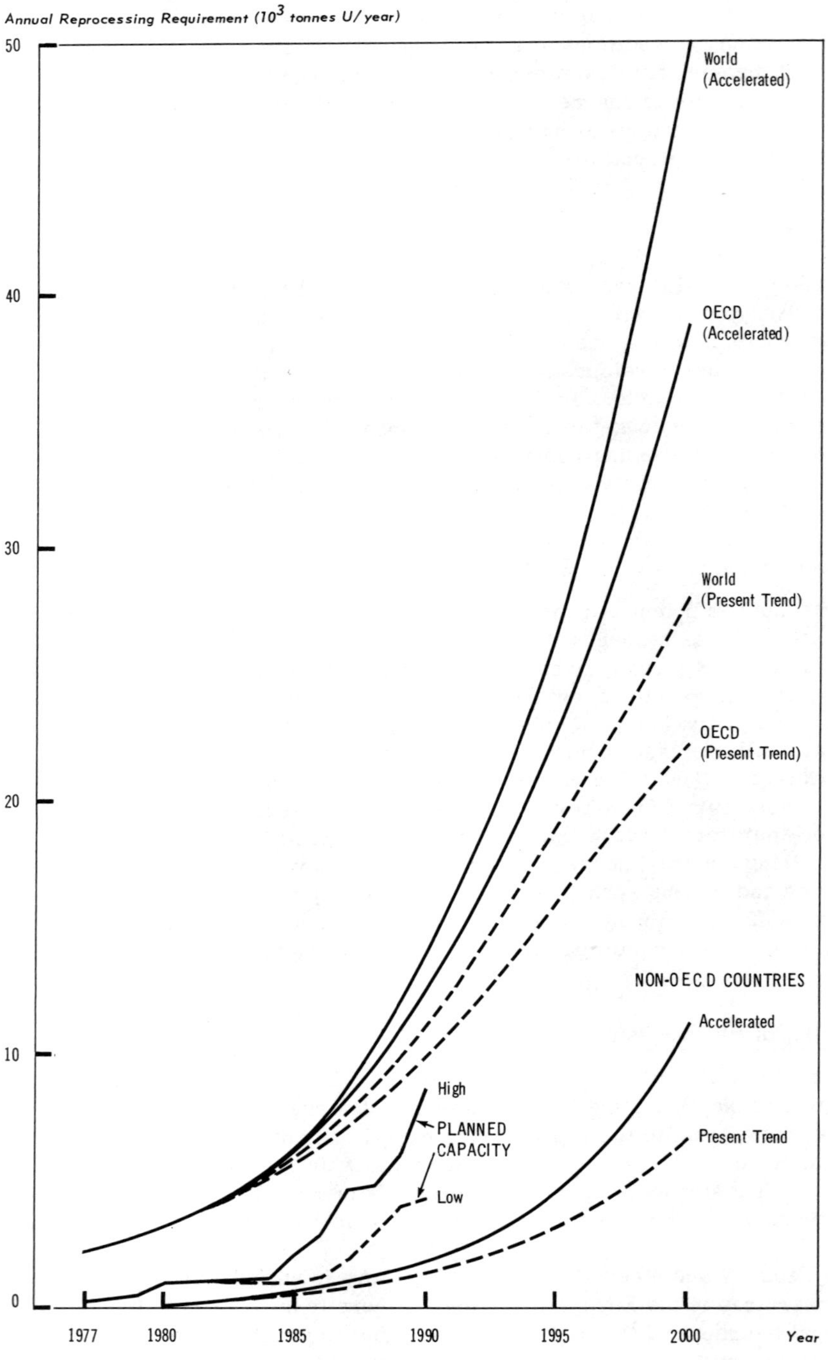

Figure 10

CUMULATIVE REPROCESSING REQUIREMENT
1977-2000

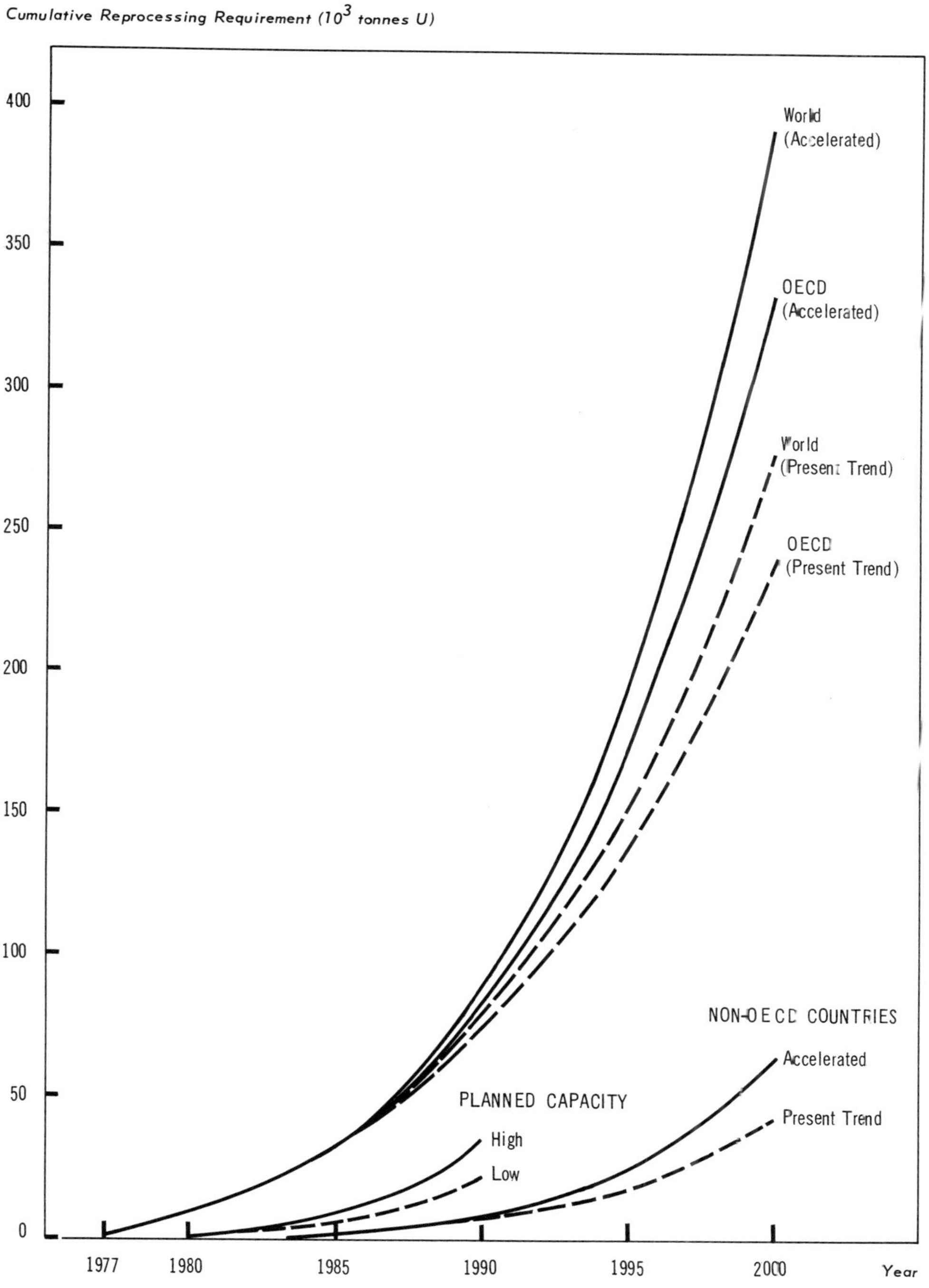

Table 15. WORLD SPENT FUEL ARISINGS, 1977-2000

1,000 tonnes U

YEAR	"ACCELERATED" POWER GROWTH		"PRESENT TREND" POWER GROWHT	
	ANNUAL	CUMULATIVE	ANNUAL	CUMULATIVE
1977	2.2	2.2	2.2	2.2
1978	2.4	4.6	2.4	4.6
1979	2.7	7.3	2.7	7.3
1980	3.1	10.4	3.1	10.4
1981	3.5	13.9	3.5	13.9
1982	3.9	17.8	3.9	17.8
1983	4.6	22.4	4.5	22.3
1984	5.4	27.8	5.2	27.5
1985	6.5	34.3	5.9	33.4
1986	7.5	41.8	6.5	39.9
1987	9.1	50.9	7.4	47.3
1988	11	62	8.3	55.6
1989	12	74	9.3	64.9
1990	14	88	11	76
1991	16	104	12	88
1992	18	122	13	101
1993	20	142	15	116
1994	23	165	17	133
1995	26	191	18	151
1996	30	221	20	171
1997	34	255	22	193
1998	38	293	24	217
1999	43	336	25	242
2000	48	384	27	269

Table 16. COMPARISON OF POSSIBLE SUPPLY AND DEMAND FOR REPROCESSING SERVICES IN EUROPE IN THE SHORT-TERM

1,000 tonnes U

	1980	1985	1990
European Demand*			
Annual	0.8	2.7-3.1	4.7-6.5
Cumulative	2.3	12.0-13.5	31.7-39.3
Reprocessing Capacity**			
Annual	0.8-0.9	0.9-1.9	4.1-7.6
Cumulative	1.5-2.0	5.6-7.6	18.8-32.5

* Demand created by spent fuels from light water reactors taken from references /25/, /26/. Cumulative demand initiated in 1977.

** Commercial capacity for reprocessing spent oxide fuels taken from reference /25/ for OECD European Region, but with first operation of Windscale delayed until 1987. Cumulative capacity initiated in 1977.

Several measures have been proposed to increase the adequacy of reprocessing services and for interim storage /25, 27/, and there is some evidence that such measures are being taken.

The need for reprocessing to provide plutonium for fast breeder reactors has been stated by the European Community /9/, and there will be greater discussion of this point in the section of this report that is devoted to the long-term. There is the opinion in some other communities that plutonium held in stored spent fuels still offers the option to proceed with breeder programmes in the longer-term. However, the opposite view is that reprocessing and recycle must take place in the nearer-term not only to provide start-up plutonium for fast breeders, but to gain the experience required to achieve a reasonably short doubling time and thereby sustain a rapid and significant growth rate for nuclear power.

4. Fuel Fabrication Services

a) Demand

The annual short-term world demand for fabrication of nuclear fuel is shown on Figure 11, and listed on Table 16. The separate demands of OECD and non-OECD countries are shown on Figure 12.

Table 16. WORLD FUEL FABRICATION REQUIREMENTS, 1977-2000

1,000 tonnes U

YEAR	"ACCELERATED" POWER GROWTH		"PRESENT TREND" POWER GROWTH	
	ANNUAL	CUMULATIVE	ANNUAL	CUMULATIVE
1977	4	4	4	4
1978	5	9	5	9
1979	6	15	6	15
1980	7	22	6	21
1981	9	31	7	28
1982	10	41	8	36
1983	12	53	9	45
1984	13	66	10	55
1985	15	81	11	66
1986	17	98	13	79
1987	19	107	14	93
1988	22	129	16	109
1989	25	154	17	126
1990	28	182	19	145
1991	31	213	21	166
1992	35	248	22	188
1993	39	287	24	212
1994	43	330	26	238
1995	48	378	28	266
1996	53	431	30	296
1997	57	488	32	328
1998	63	551	34	362
1999	70	621	37	399
2000	74	695	38	437

b) Supply

Although current world capacity is adequate to fulfil needs for fabrication of fuel assemblies and for all intermediate steps down to chemical conversion, Figure 11 indicates that an increase in annual capacity of at least an order of magnitude will be required by the year 2000. The acknowledged importance

Figure 11

WORLD ANNUAL FUEL FABRICATION REQUIREMENTS
1977-2000

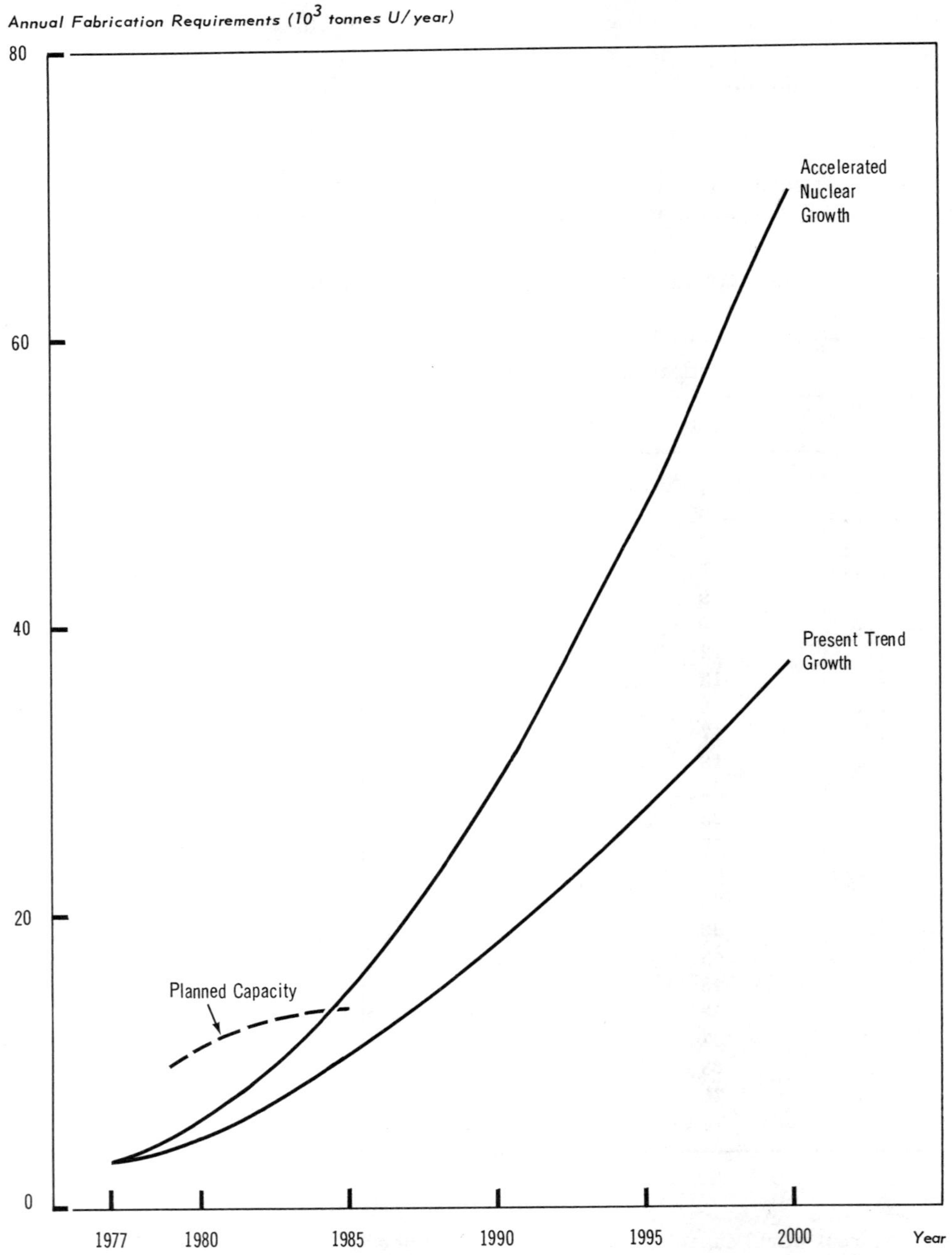

Figure 12

ANNUAL FUEL FABRICATION REQUIREMENTS
1977-2000
OECD and Non-OECD Countries

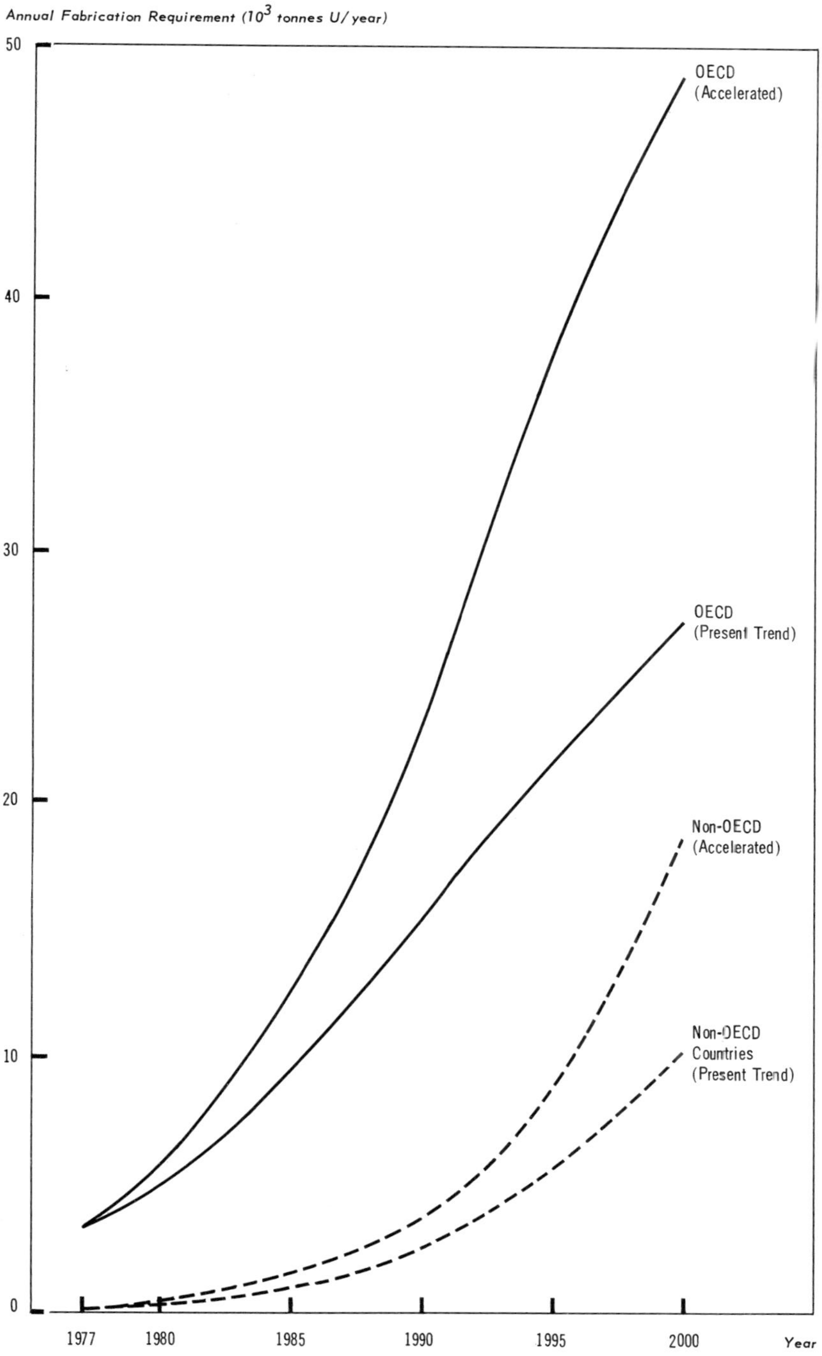

of this industry will undoubtedly be an important factor in attaining the necessary growth. The capacity planned by the industry through 1985 /26/ has been plotted on Figure 11, and appears to be adequate through that date.

Aside from coping with the plysical problem of expansion, the industry must also deal with technological changes, due to the introduction of breeder fuels or of other advanced reactor fuels that may be introduced in the latter part of the short-term.

V

THE LONGER-TERM (2000-2025)

Because the short-term outlook for nuclear power is itself fraught with uncertainty, longer-term forecasts have of late fallen into disfavour. However, as many plans and decisions that must be taken today depend on future trends in nuclear power, the adequacy of uranium resources, and possible demands for fuel cycle services, it is necessary to cope with the higher order of uncertainty inherent in longer-range estimates.

A. NUCLEAR POWER GROWTH ESTIMATES

The short-term nuclear power estimates, extending to the year 2000, provide the basis for further extension into the longer-term. The same methodology has been utilized to project nuclear power growth to the year 2025; i.e.: estimation of total primary energy requirements; electrical power's share of total energy; and the extent to which nuclear energy will replace conventional fuels in the electrical power market. These estimated parameters, together with derived total electric and nuclear electric capacities, are shown in Appendix II.* It should be noted that the two estimates of nuclear capacity are now considered as extensions into the longer term of "present trend" and "accelerated" estimates.

Table 17 shows a comparison of estimates of world primary energy needs and of installed nuclear power capacity in the year 2025.

Table 17. COMPARISON OF ESTIMATES OF THE WORLD'S PRIMARY ENERGY NEEDS AND NUCLEAR CAPACITY IN THE YEAR 2025

	WORLD PRIMARY ENERGY, EJ (10^{18} joules)	NUCLEAR ELECTRICAL CAPACITY GW(e)
This Report**	550/1,088	2,157/6,650
A. Giraud /5/	815	5,200
R.D. Vaughan, et al. /6/	494	4,500*
Häfele and Sassin /7/	945	-
M.F. Duret, et. al. /13/	400-520***	3,423***

* For the most probable breeder growth.

** For extensions of "present trend"/"accelerated" estimates.

*** For the year 2020.

The growth of nuclear power from 2000 to 2025 is plotted on Figure 13. It may be observed that the long-term extensions of the "present trend" and "accelerated" scenarios diverge rapidly, reaching 2,200 and 6,700 GWe, respectively, by the year 2025. This is characteristic of the extreme uncertainties

* - Total energy requirements: Table II-1.
- Electrical share of total energy, system load factor, and nuclear share of electrical capacity: Figure I-1.
- Total electric power requirements and capacities, and nuclear power capacities: Table II-2.

Figure 13

WORLD NUCLEAR POWER GROWTH*

2000-2025

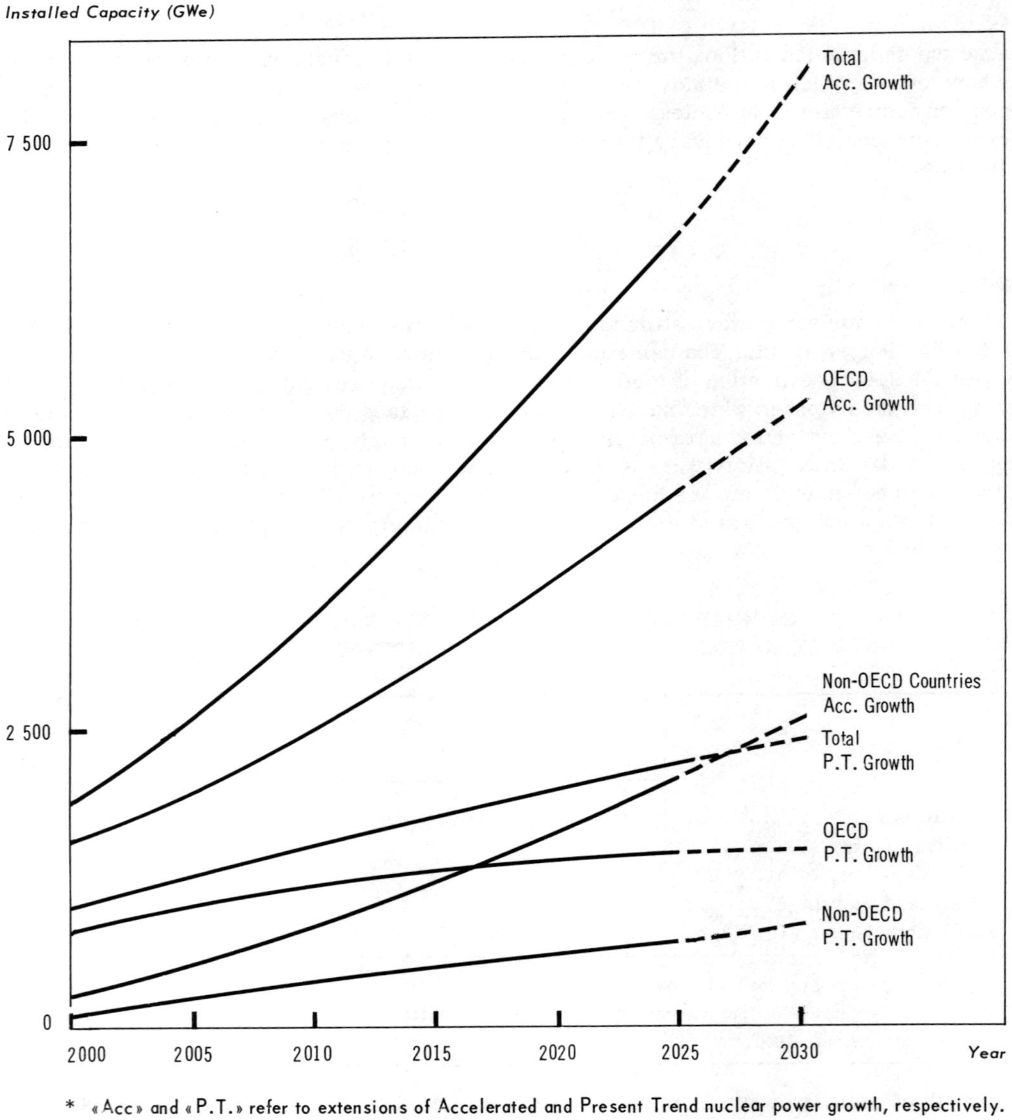

* «Acc» and «P.T.» refer to extensions of Accelerated and Present Trend nuclear power growth, respectively.

in long-range nuclear forecasting, but it should be stressed that the intent in this section of the report is to identify general relationships between reactor strategies and fuel cycle needs rather than to quantify. Within this context, the forecast serves us well.

B. LONG-TERM REACTOR STRATEGIES

Quite obviously, a large variety of reactor strategies could be postulated to meet these patterns of growth. An attempt has been made to keep the selected strategies to a reasonable number, but with a view toward encompassing a wide range of possible demands on fuel cycle services. The reactor concepts chosen are those of current commercial interest, and their characteristics are listed, together with those used in the shorter-term, in Table I-10, Appendix I.

It should also be stated that with the exception of the LWR strategy, all of the chosen reactor strategies require spent fuel reprocessing and recycle. This is not an unreasonable condition, as the presently clouded outlook for reprocessing and recycle should be cleared up by the long-term. Finally, it should be noted that the reactor distributions in-being at the end of the short-term, serve as the starting point for the implementation of the new strategies of the longer-term.

Five fundamental reactor strategies have been selected:

1. LWR Strategy

Here, it is assumed that the nuclear power market will continue to be dominated by light water reactors, with no significant penetration by other reactor concepts. New LWR capacity is continually brought on-line, up to the limits of the forecasts. Although this pure strategy (and particularly the non-recycle variant) is unrealistic in the longer-term due to its large demand for uranium, it is included as a point of reference.

2. LWR + FBR Strategy

In this strategy, the introduction of the fast breeder reactor, which commences in 1995, is constrained during a 15-year transitional period to the limits of either of two specified rates (subject to the availability of plutonium) and thereafter determined either by the availability of plutonium or by the limit of the forecast of total nuclear capacity, whichever is controlling. The two FBR introduction rates imposed from 1995 until the year 2010 are: (1) for "present trend" nuclear power growth, FBR capacity roughly follows the historical pattern of LWR introduction that began in 1964; (2) for "accelerated" nuclear power growth, FBRs are introduced at a rapid rate, accounting for virtually all new European capacity after 1995. The actual FBR introduction schedules are the following:

GWe of Installed FBR Capacity

NUCLEAR POWER ESTIMATES	2000	2005	2010
Extension of "Accelerated"	145	615	1,500
Extension of "Present trend"	10	30	60

As variants in this case, "early" and "advanced" FBR models are employed, with either "rapid" or "slow" plutonium recycle, in order to obtain a wide range of doubling times.

3. LWR + HWR + FBR Strategy

This mixed strategy involves a substantial increase in the construction of heavy water reactors of CANDU type beginning in 1985 (at the historical growth rate of LWRs, dating back to 1964). The fast breeder is then introduced in 1995, as in strategy 2, using either early or advanced models of FBRs.

4. LWR + HWR (thorium cycle)

This strategy is analogous to strategy 2, except that an advanced heavy water reactor, utilising the thorium cycle, is introduced in 1995 rather than the fast breeder.

5. LWR + HTR

Again, this strategy is a variant of strategy 2, except that the high-temperature, gas-cooled reactor is introduced in 1995, rather than the fast breeder. As a variant, either highly-enriched or slightly-enriched HTRs are utilized.

The reactor distributions that were computed in accord with the conditions of each of these strategies (with the exception of the pure LWR strategy) are shown as Figures II-1 through II-4, Appendix II.

These diverse scenarios can be expected to generate a wide range of demands for uranium and fuel cycle services, which can now be examined and compared.

C. URANIUM REQUIREMENTS AND SUPPLY

1. Requirements

Figures 14 through 22 are graphs of the demands for uranium under each of the reactor strategies, as follows:

Strategie 1 (LWR):

Figure 14 - Annual demand
Figure 15 - Effects of forecast parameters on annual demand
Figure 16 - Cumulative demand

Strategie 2 (LWR + FBR):

Figure 17 - Annual demand
Figure 18 - Cumulative demand

Strategy 3 (LWR + HWR + FBR):

Figure 19 - Annual demand
Figure 20 - Cumulative demand

Strategy 4 (LWR + HWR - Th cycle), and

Strategy 5 (LWR + HTR):

Figure 21 - Annual demand
Figure 22 - Cumulative demand

On Table 19, the annual and cumulative uranium requirements by the year 2025 for all of the reactor strategies are compared.

Although a detailed examination must be made of each Figure to appreciate the full implications of these requirements, some general observations can be made. First, it is clear that in the long-term, a purely light-water strategy is predictably wasteful of uranium resources. For example, minimum LWR requirements, assuming a continuation of the "present trend" in the growth of nuclear power, will be 6 million tonnes of uranium by the year 2025, assuming full recycle in the longer-term. A continuation of the short-term inability to recycle would increase this figure to 9 million tonnes of uranium, while all requirements of the "accelerated" nuclear economy are so high as to be unfeasible on the basis of current estimates of resources. Moreover, LWR uranium requirements could double during 2025-2050.

In several of the advanced-converter strategies, particularly that of the HWR utilizing the thorium cycle, and to a lesser extent, the HTRs, considerable reductions in uranium demands may be realized, when compared to the pure LWR strategy. A detailed study of the prospects for such reactors is being prepared /28/. However, although advanced converters utilize uranium more efficiently than light water reactors, their cumulative demands for uranium, particularly in a high nuclear power growth scenario, still far exceed known resources, while their annual demands increase continuously. Further, the more attractive advanced converters create an additional demand for thorium. As an example, the

Figure 14

WORLD ANNUAL URANIUM DEMAND
2000-2025

LWR Strategy

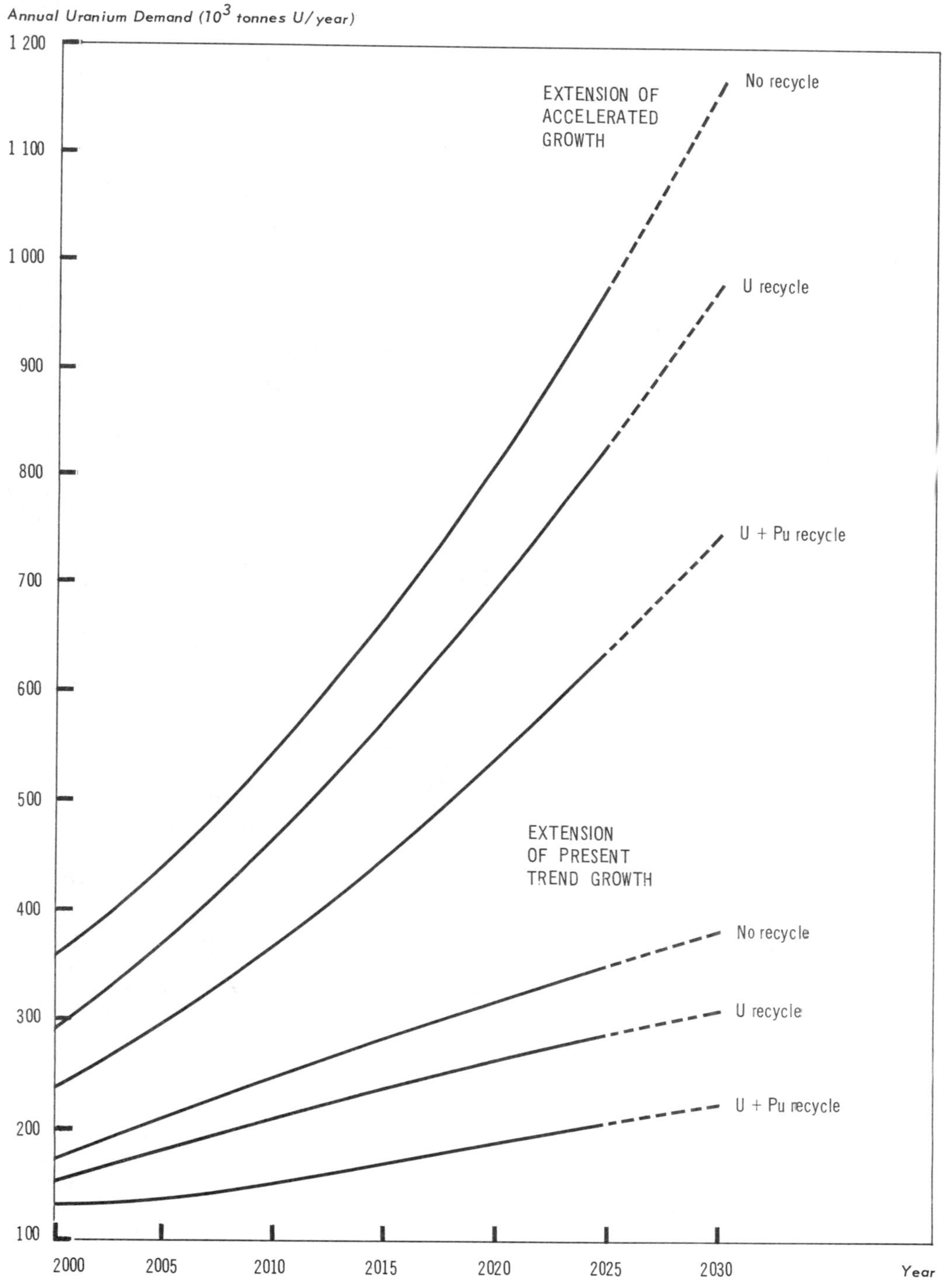

Figure 15

EFFECTS OF PARAMETERS ON WORLD ANNUAL DEMAND FOR URANIUM

LWR Strategy

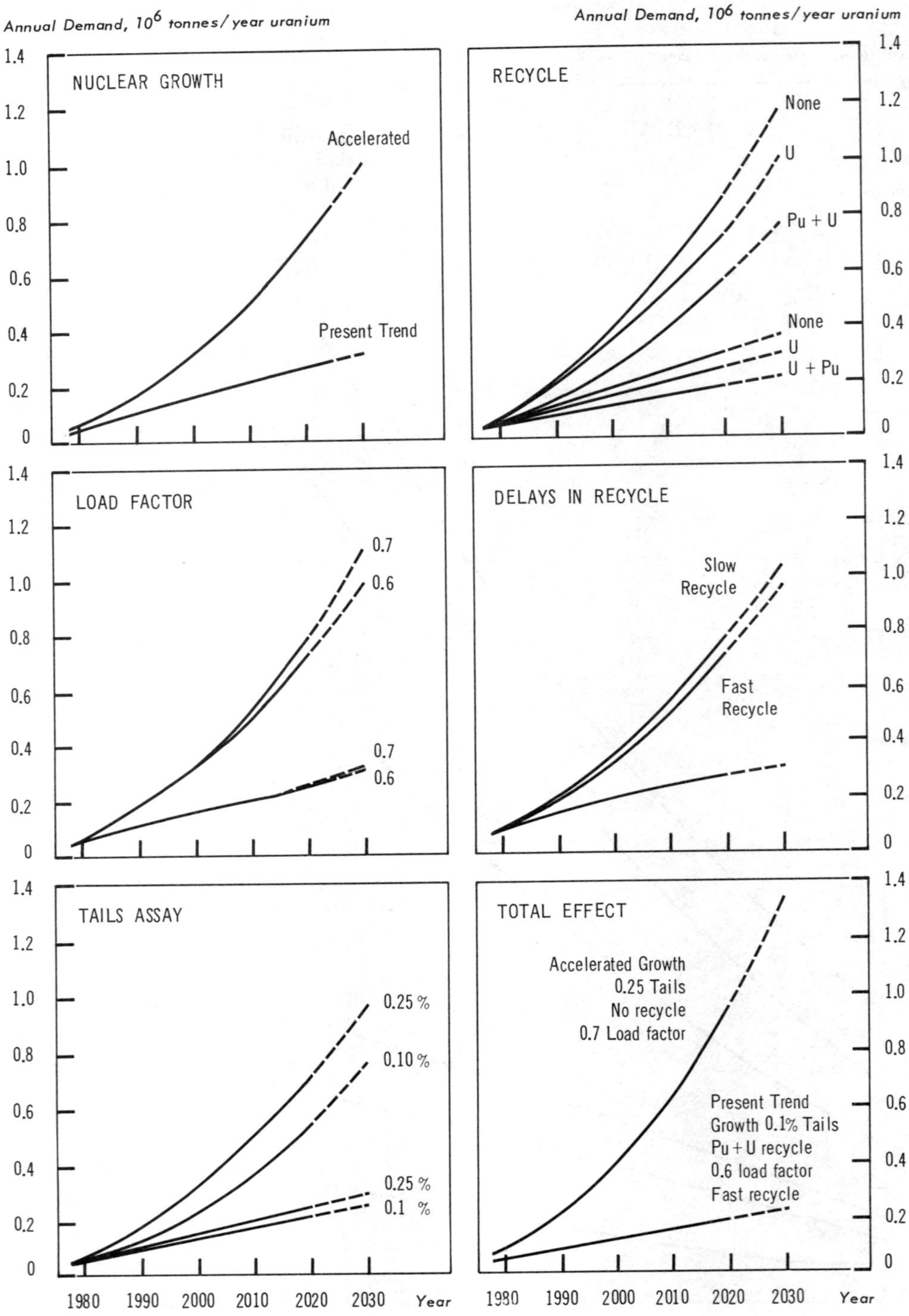

Figure 16

WORLD CUMULATIVE URANIUM DEMAND*
2000-2025

LWR Strategy

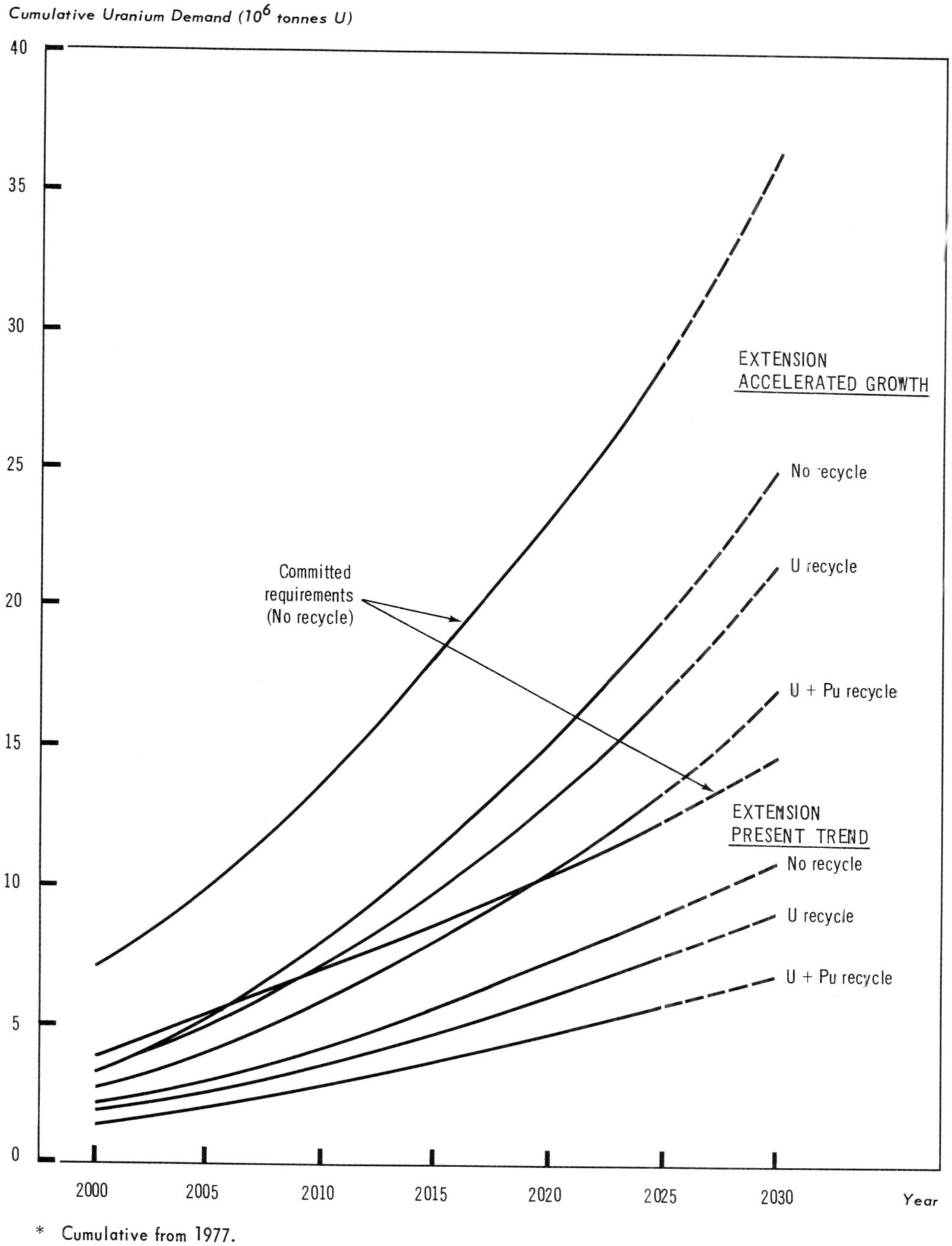

* Cumulative from 1977.

Figure 17

WORLD ANNUAL URANIUM DEMAND*

2000-2025

LWR + FBR Strategy

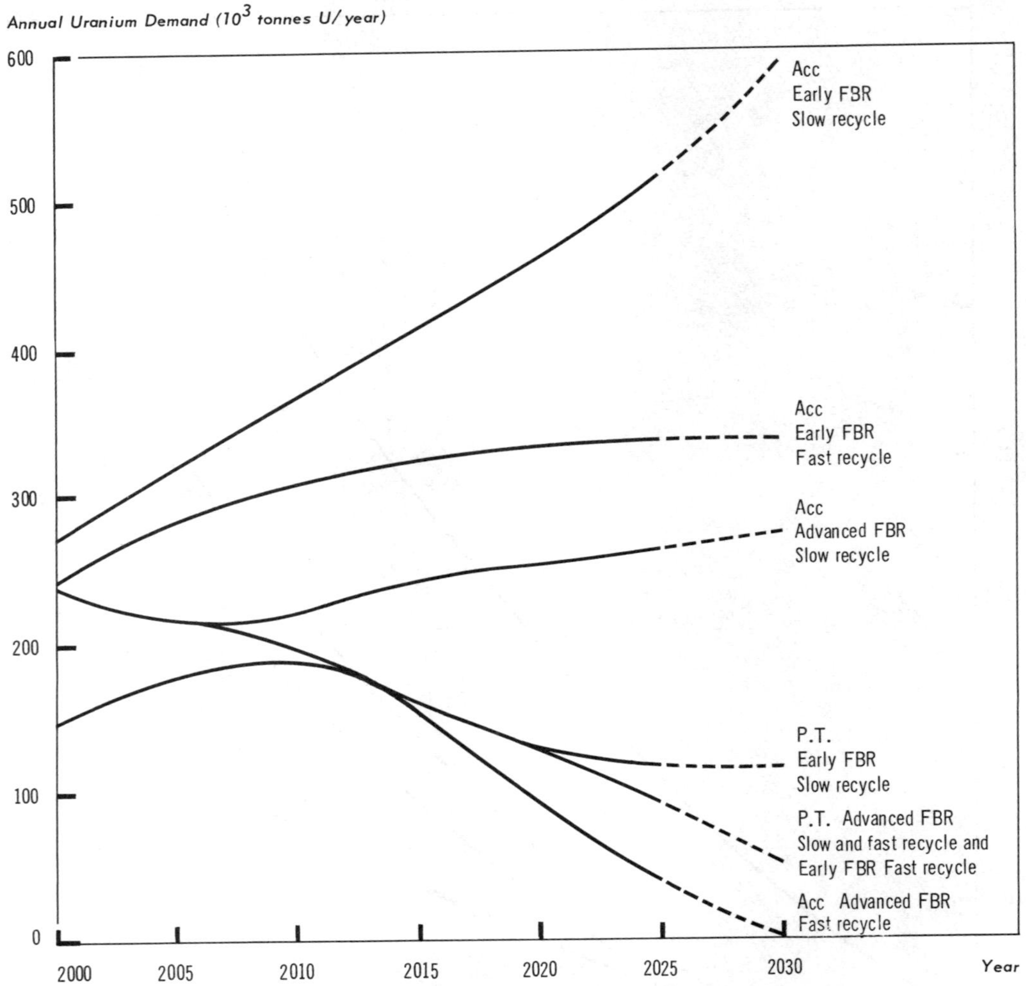

* «Acc» and «P.T.» refer to extensions of Accelerated and Present Trend nuclear power growth, respectively.

Figure 18

WORLD CUMULATIVE URANIUM DEMAND*
2000-2025

LWR + FBR Strategy

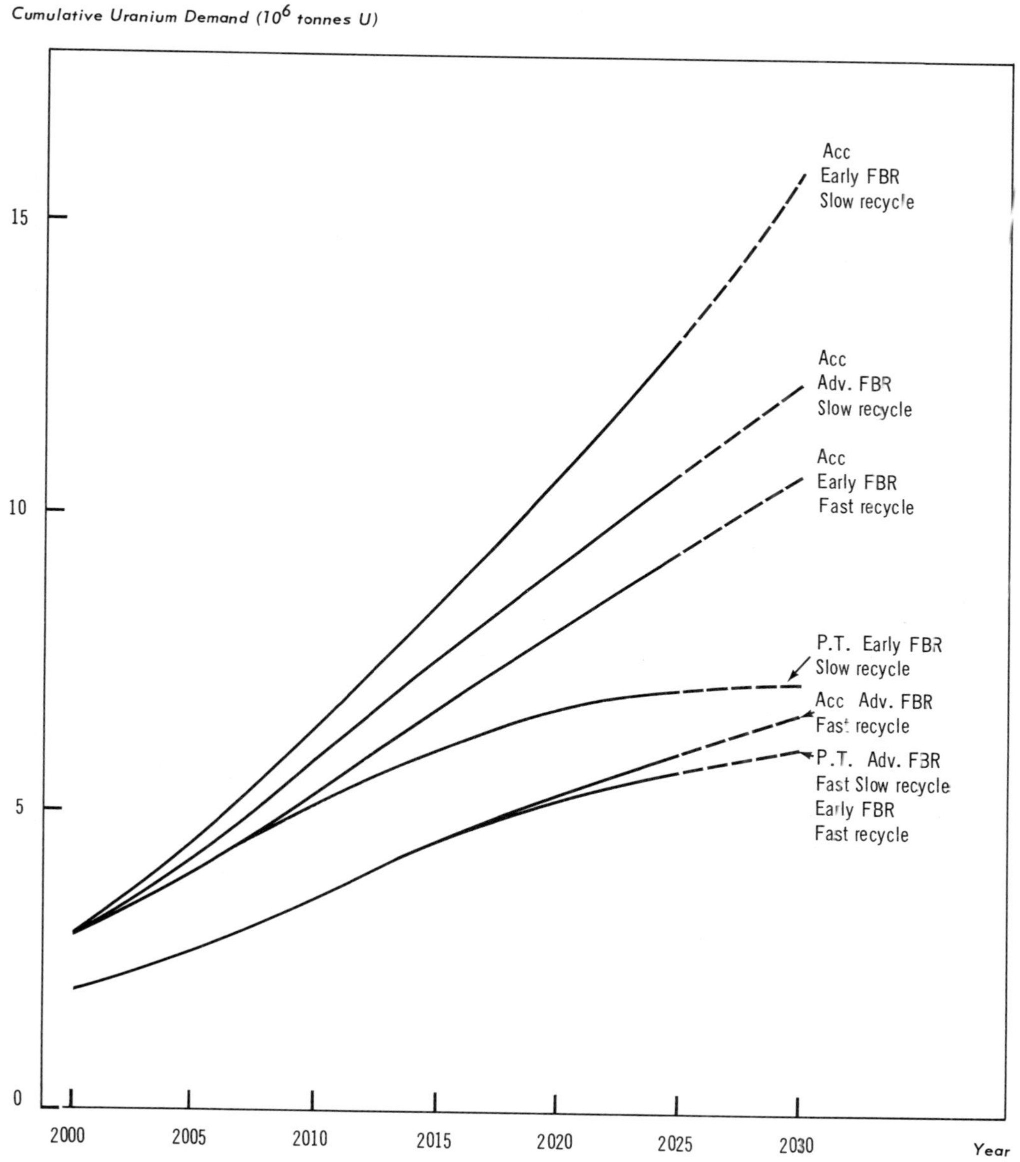

* Cumulative from 1977.
«Acc» and «P.T.» refer to extensions of Accelerated and Present Trend nuclear power growth, respectively.

Figure 19

WORLD ANNUAL URANIUM DEMAND*
2000-2025
LWR + HWR + FBR Strategy

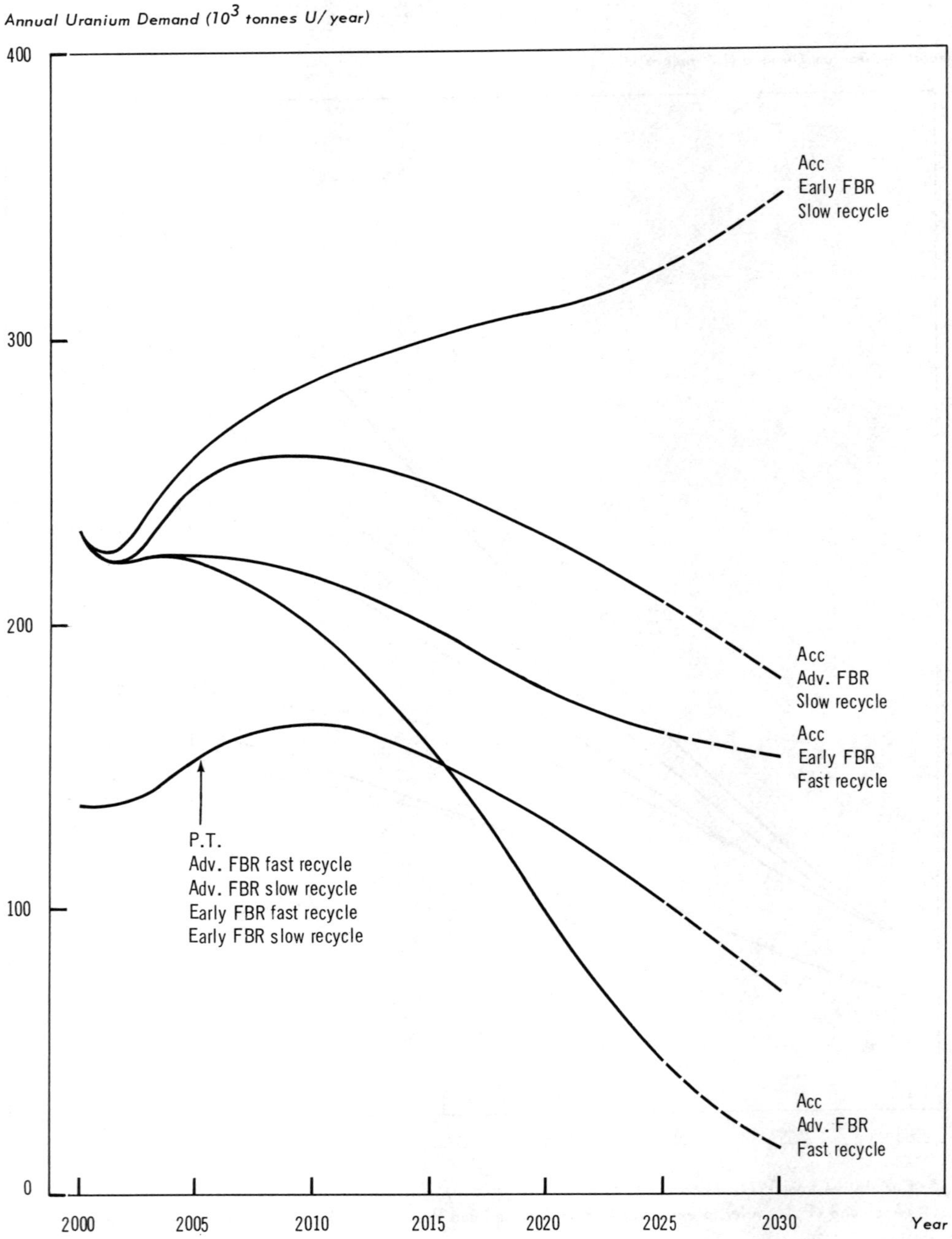

* «Acc» and «P.T.» refer to extensions of Accelerated and Present Trend nuclear power growth, respectively.

Figure 20

WORLD CUMULATIVE URANIUM DEMAND*
2000-2025

LWR + HWR + FBR Strategy

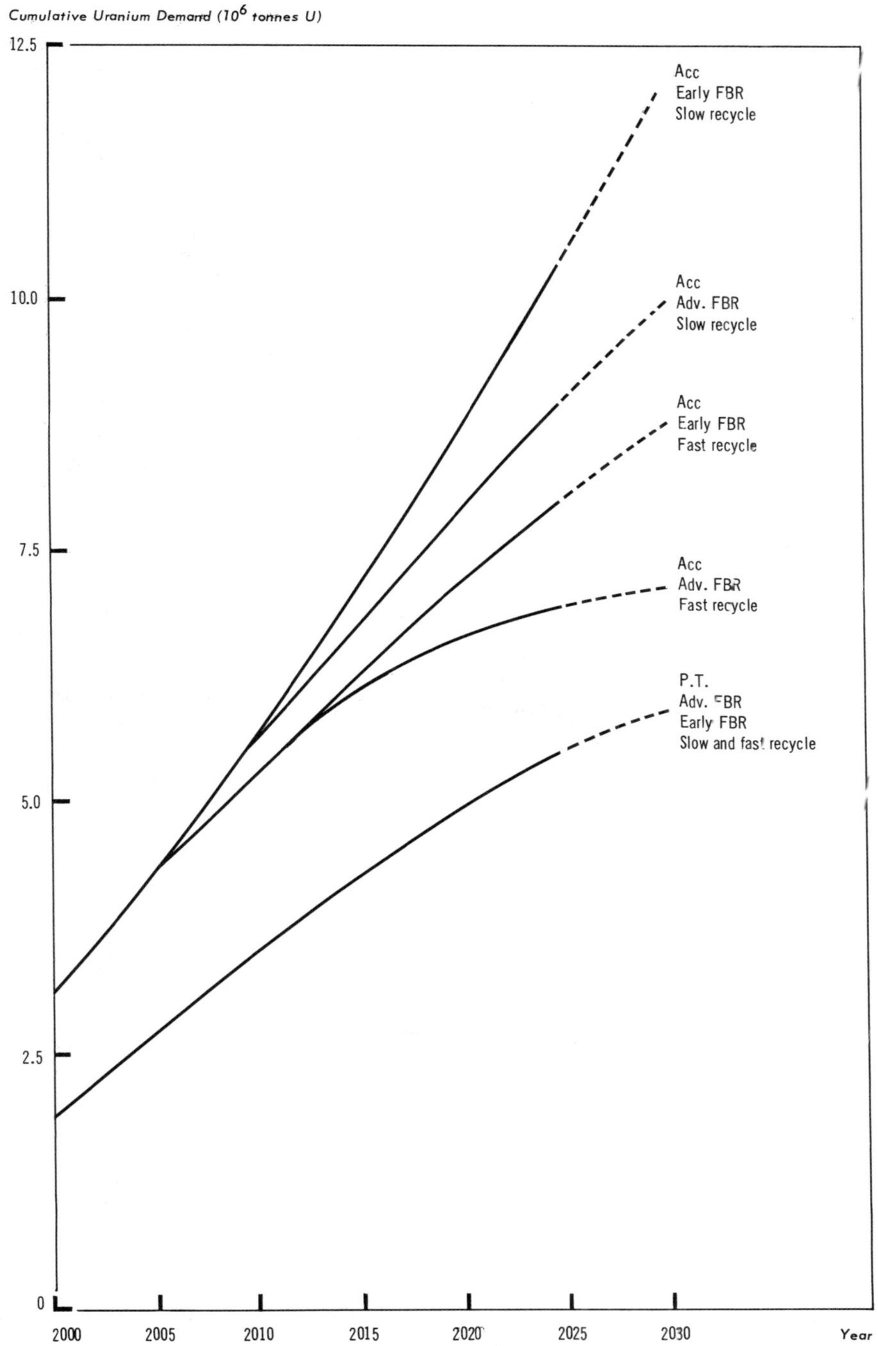

* Cumulative from 1977.
«Acc» and «P.T.» refer to extensions of Accelerated and Present Trend nuclear power growth, respectively.

Figure 21

WORLD ANNUAL URANIUM DEMAND* FOR LONG TERM
2000-2025

LWR + HTR and LWR + HWR strategies

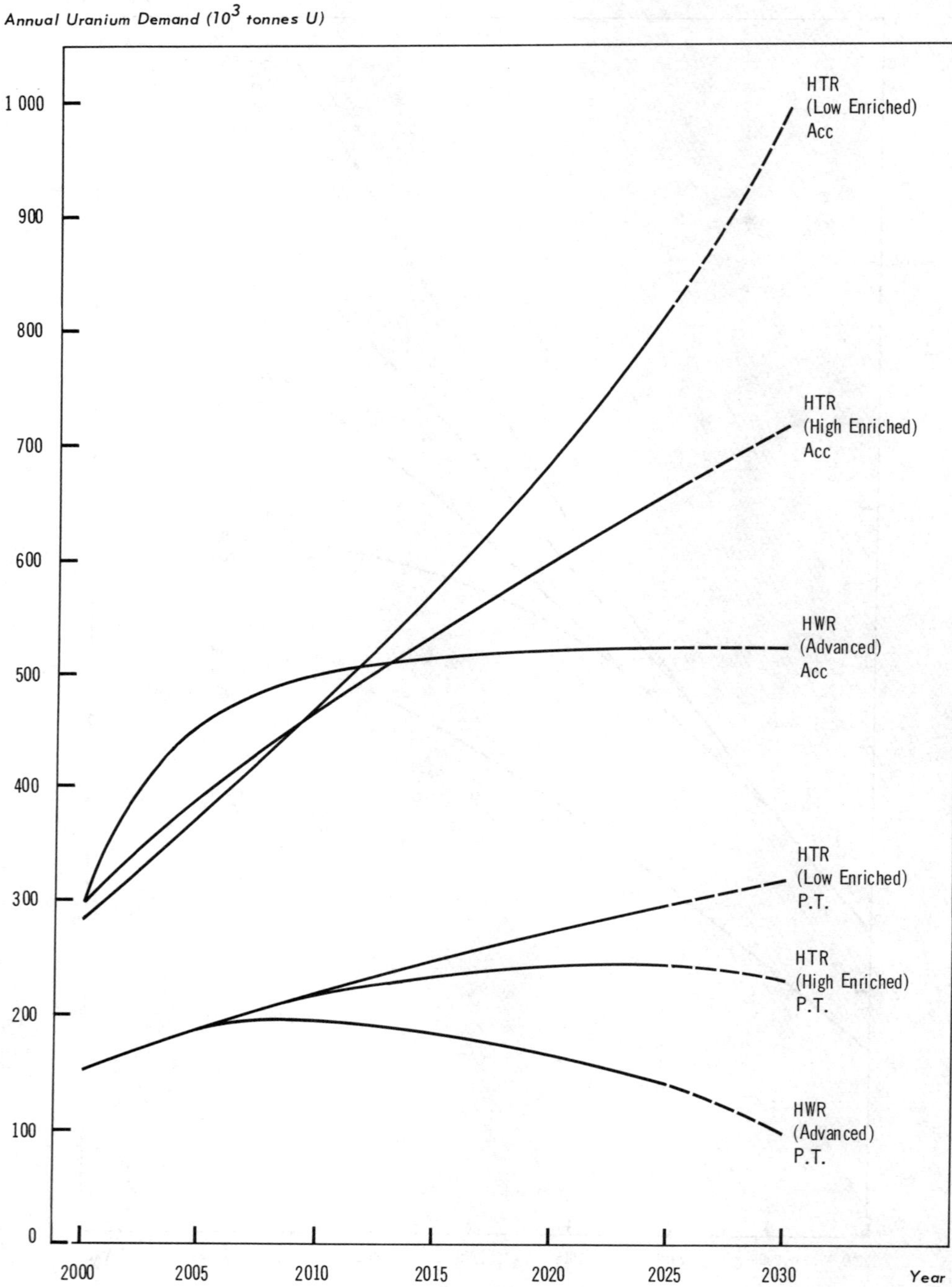

* «Acc» and «P.T.» refer to extensions of Accelerated and Present Trend nuclear power growth, respectively.

Figure 22

WORLD CUMULATIVE URANIUM DEMAND*

2000-2025

LWR + HTR and LWR + HWR Strategy

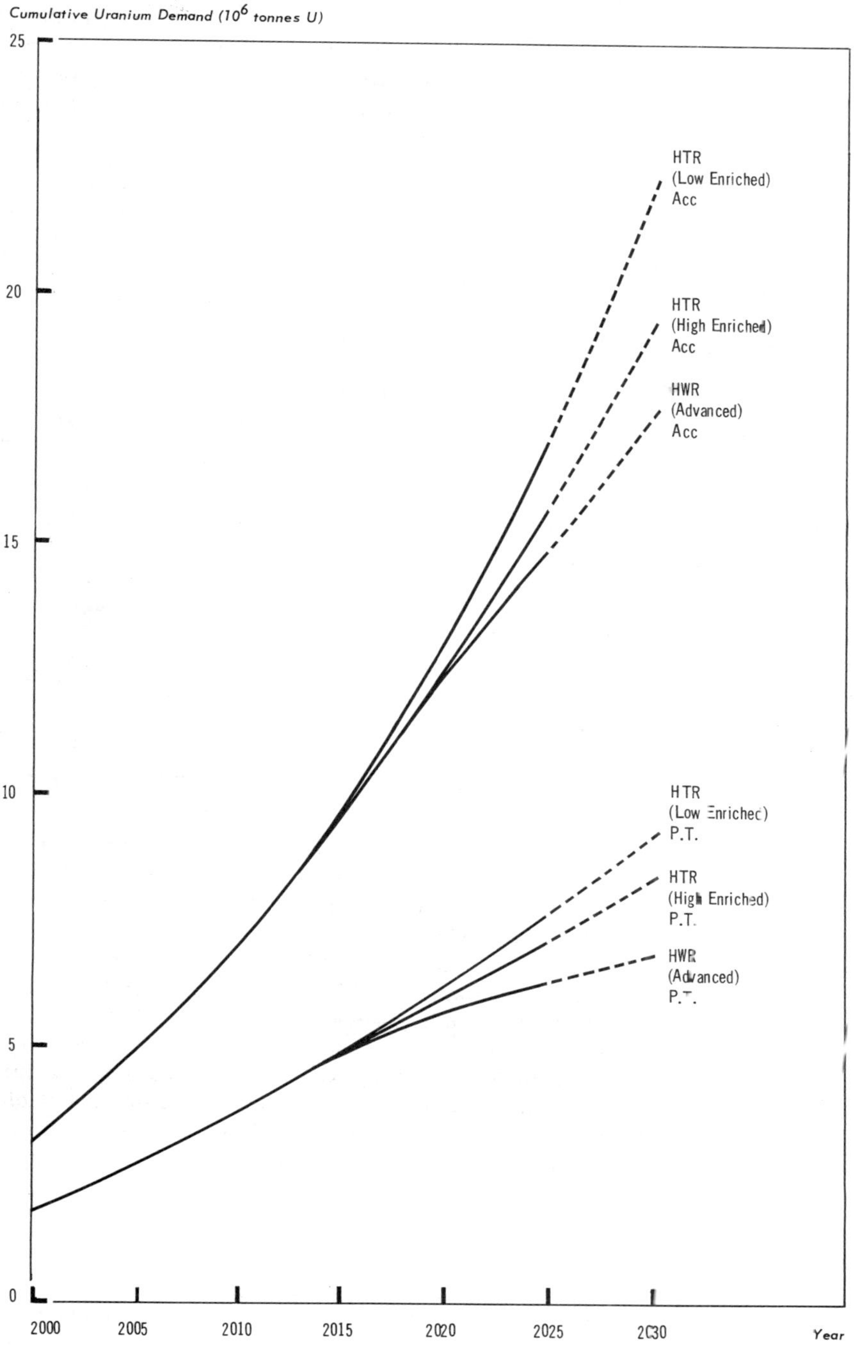

* Cumulative from 1977.

«Acc» and «P.T.» refer to extensions of Accelerated and Present Trend nuclear power growth, respectively.

Table 19. COMPARISON OF WORLD REQUIREMENTS FOR URANIUM BY THE YEAR 2025, UNDER ALTERNATIVE STRATEGIES

REACTOR STRATEGY IN LONG-TERM		ANNUAL REQUIREMENT (THOUSAND TONNES U/YEAR)		CUMULATIVE REQUIREMENT, 1977-2025 (MILLION TONNES U)	
STRATEGY		NUCLEAR POWER GROWTH - EXTENSION OF:		NUCLEAR POWER GROWTH - EXTENSION OF:	
No.	DESCRIPTION	"PRESENT TREND"	"ACCELERATED"	"PRESENT TREND"	"ACCELERATED"
1.	LWRs only, once-through fuelling	350	990	9.0	20
	LWRs only, spent fuel recycle	210	640	5.8	13
2.	LWR + advanced FBR, fast recycle ...	95	40	5.8	7.1
	LWR + advanced FBR, slow recycle ..	95	310	5.8	10.8
	LWR + early FBR, fast recycle	95	260	5.8	9.3
	LWR + early FBR, slow recycle	120	520	6.0	13.0
3.	LWR + HWR + FBR				
	advanced FBR, fast recycle	100	40	5.5	7.0
	advanced FBR, slow recycle	100	210	5.5	9.0
	early FBR, fast recycle	100	160	5.5	8.0
	early FBR, slow recycle	100	320	5.5	10.5
4.	LWR + HWR (thorium cycle)	145	540	6.4	15.5
5.	LWR + HTR (High enrichment)	240	660	7.4	16
	LWR + HTR (Low enrichment)	290	840	7.6	17

cumulative demand for thorium by the year 2025 would be 0.07-0.3 million tonnes for the HTR (highly-enriched) and 0.6-2.0 million tonnes for the HWR (Th cycle) strategies. World thorium resources, largely undeveloped, are now estimated as 0.49 million tonnes in the "Reasonably Assured" category and 0.83 million tonnes in the "Estimated Additional" category /2/.

On the other end of the scale, the impact of the fast breeder reactor on uranium demand is impressive. One can envision cumulative uranium demand leveling out at slightly more than 6 million tonnes in the post 2025 period if doubling time can be reduced to the order of 25-30 years or less. Perhaps of even greater significance is the fact that the annual requirements curve in most FBR cases goes through a maximum before the year 2025, indicating extension of remaining uranium resources well beyond that date. The large depleted uranium inventory on hand in 2025 would be sufficient to sustain the breeders' fueling requirements for many hundreds of years. Thus, the breeder if demonstrated to be economically feasible, can indeed be regarded as an "inexhaustible source of energy supply".

2. Supply Prospects

As the uranium requirements in the longer-term involve large quantities that are beyond known resources levels, any discussion of the prospects for supply must necessarily include some speculation on the success of future exploration efforts. Sharp disparities between past prospecting efforts have led to a heavy concentration of low cost uranium reserves in half a dozen countries, and this situation cannot be assumed to continue indefinitely. Very large areas of the world remain to be explored and there is a probability that much more uranium remains to be discovered in the upper part of the earth's crust. Purely statistical attempts to estimate world uranium resources have resulted in figures upwards of several tens of millions of tonnes uranium. As noted in reference /29/, such methods have their limitations, and greater credence must be placed in estimates based on reasonable extensions of known geological data. In the latter category, a figure of 8.8 million tonnes uranium has been ventured as a world speculative potential /30/, in a cost range that now includes material up to $ 130/kg U ($ 50/lb U_3O_8). As there is a range of technical opinion as to the amount of uranium that can presently be considered economically recoverable, the wisest course will be to expand exploration and exploration technology, in order to support the necessary production effort /31/.

With respect to higher cost uranium resources, relatively few countries have attempted to quantify resources in this category. The problem to be considered is how much of this material can be made available at a cost acceptable to the nuclear power industry within a useful time period.

Finally, it should be added that the incentives necessary to insure an adequate supply of uranium, as discussed earlier with respect to short-term demand, are of equal significance in the long-term.

D. ENRICHMENT

1. Demand

Enrichment demands of each of the reactor strategies are illustrated on Figures 23 and 24.

It may be observed that demands for enriching services continue to increase steadily throughout the longer-term, except for some strategies involving the FBR and HWR (thorium cycle). In the latter instance, it could even be possible to reduce enrichment demands to the order of present levels by the year 2025.

2. Supply Outlook

It is difficult to speculate about the provision of enrichment services over the long-term. Both the positive and negative factors discussed in the section on short-term supply will still pertain, although it is expected that prospective enrichers will be looking for early signs of a downturn in the market due to the advent of breeders or other advanced reactors. It is also possible that new enriching methods may be developed that could make it more economically attractive to reach lower tails assays and thereby reduce uranium requirements.

E. REPROCESSING SERVICES

1. Demand

It is to be noted that all reactor strategies considered in the longer-term require spent fuel reprocessing and recycle unless major new uranium resources well beyond current best estimates are discovered and developed. Thus, if nuclear power is to play a significant role in furnishing the long-term energy needs of mankind, one must now speak in terms of "reprocessing demand" rather than "spent fuel arisings".

The reprocessing demand for all reactor strategies, is shown on Figures 25 and 26.

2. Supply

Although the longer-term outlook for the supply of reprocessing services is obscured by the present difficulties in the short-term, it is possible, with the passage of time and the introduction of advanced reactors requiring this essential services, that the present impasse will be resolved. It is

Figure 23

WORLD ANNUAL ENRICHMENT DEMAND
2000-2025

Extension of Accelerated Nuclear Power Growth

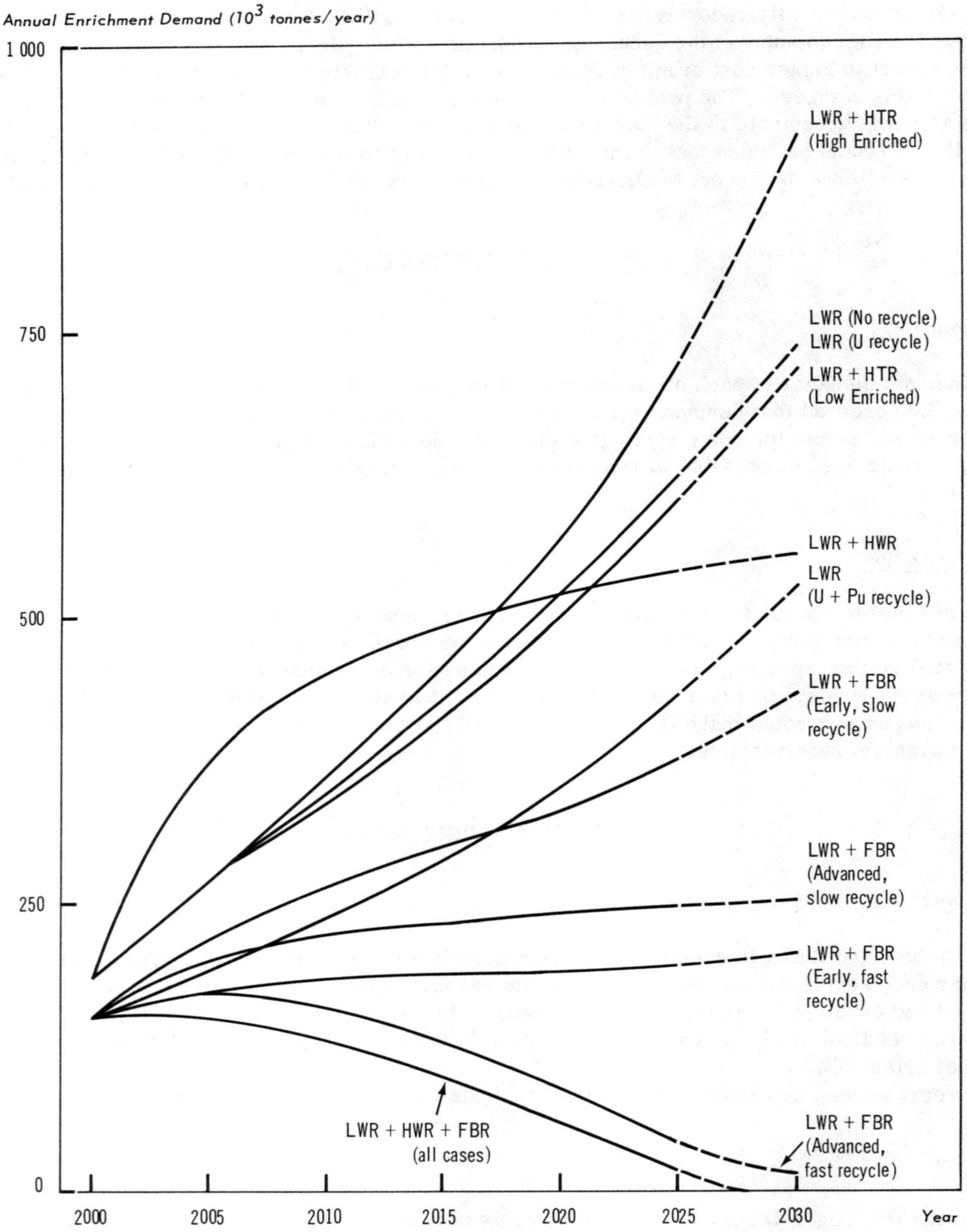

Figure 24

WORLD ANNUAL ENRICHMENT DEMAND
2000-2025

Extension of Present Trend Nuclear Power Growth

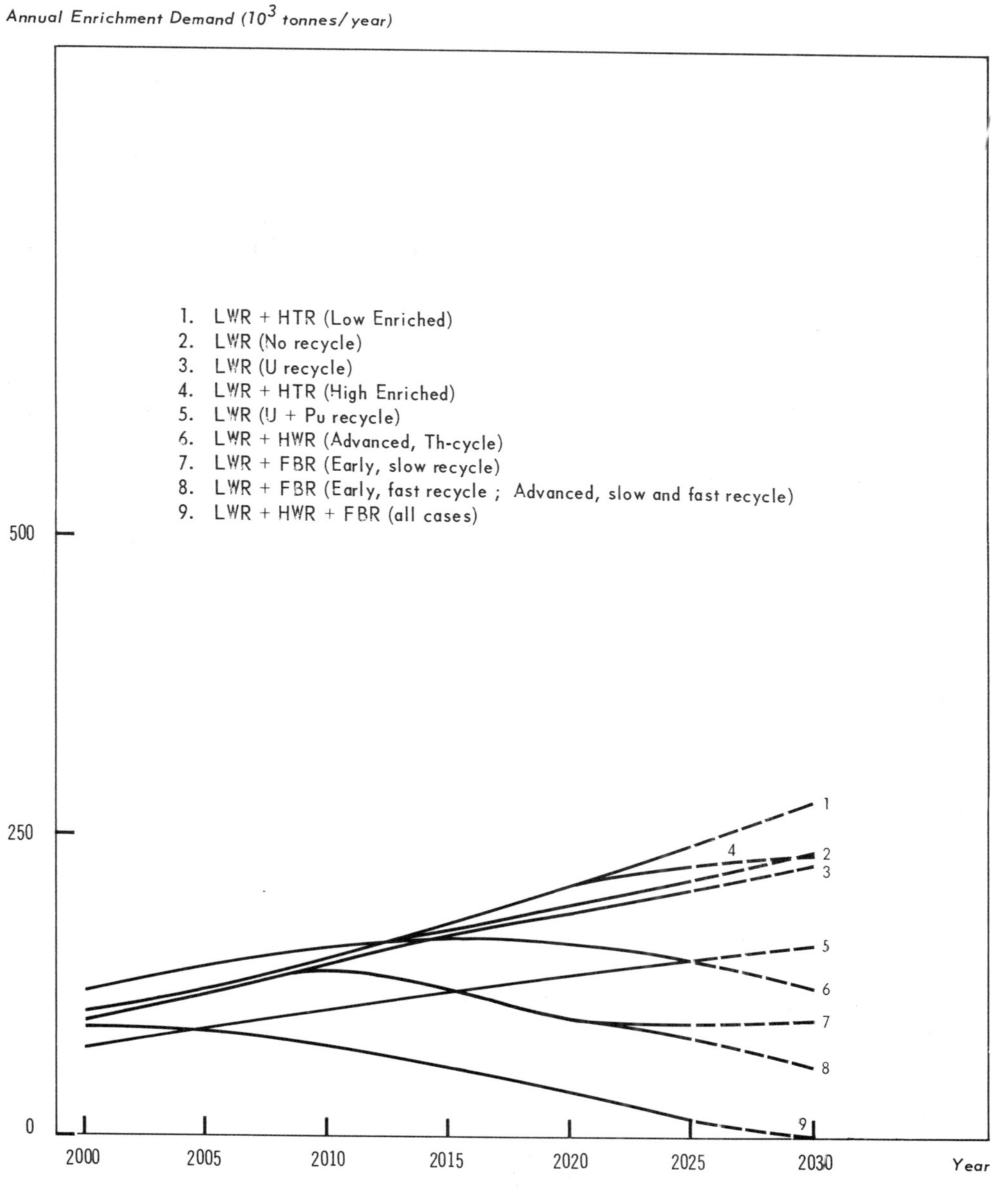

Figure 25

WORLD ANNUAL REPROCESSING DEMAND
2000-2025

Extension of Accelerated Nuclear Power Growth

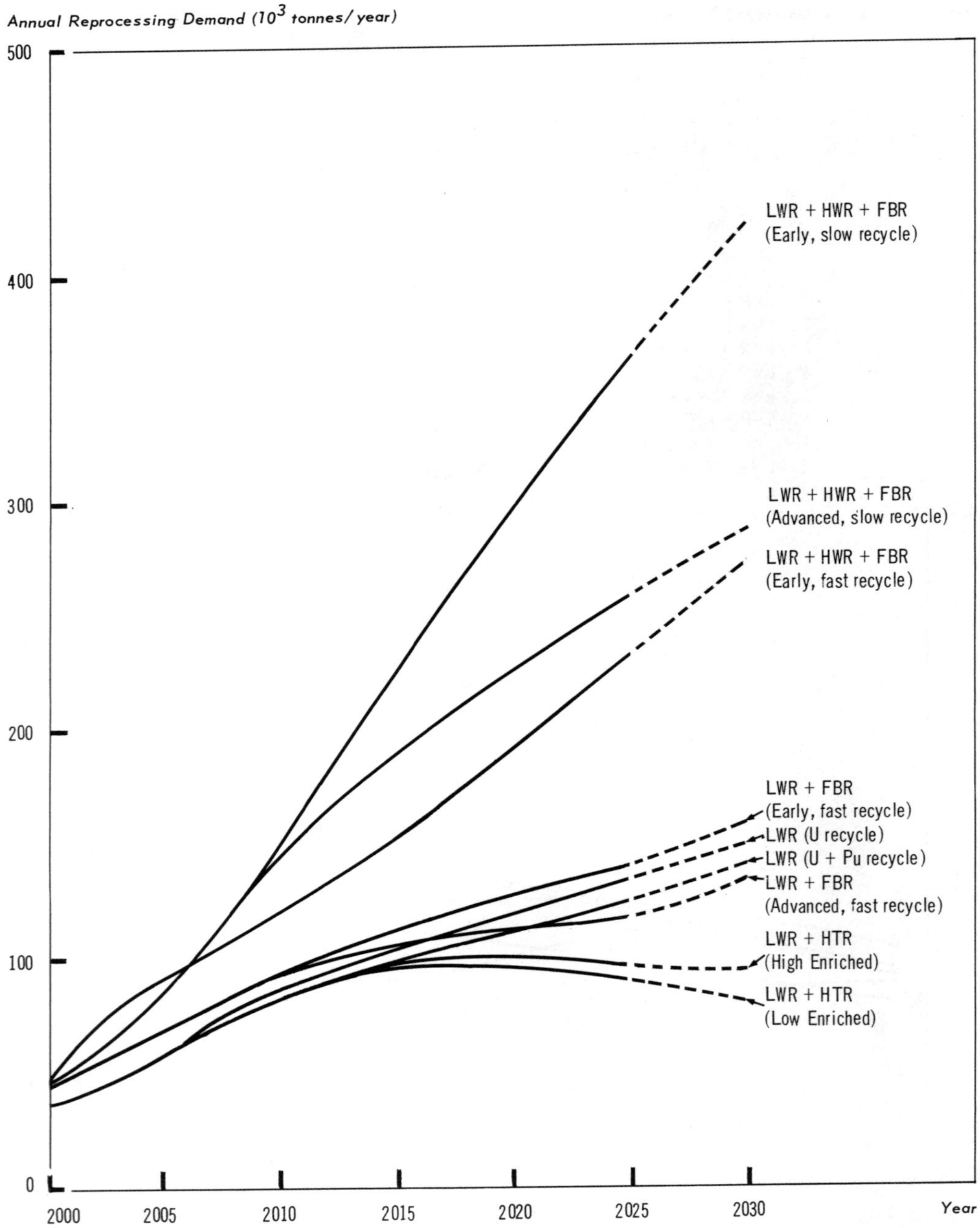

Figure 26

WORLD ANNUAL REPROCESSING DEMAND
2000-2025

Extension of Present Trend Nuclear Power Growth

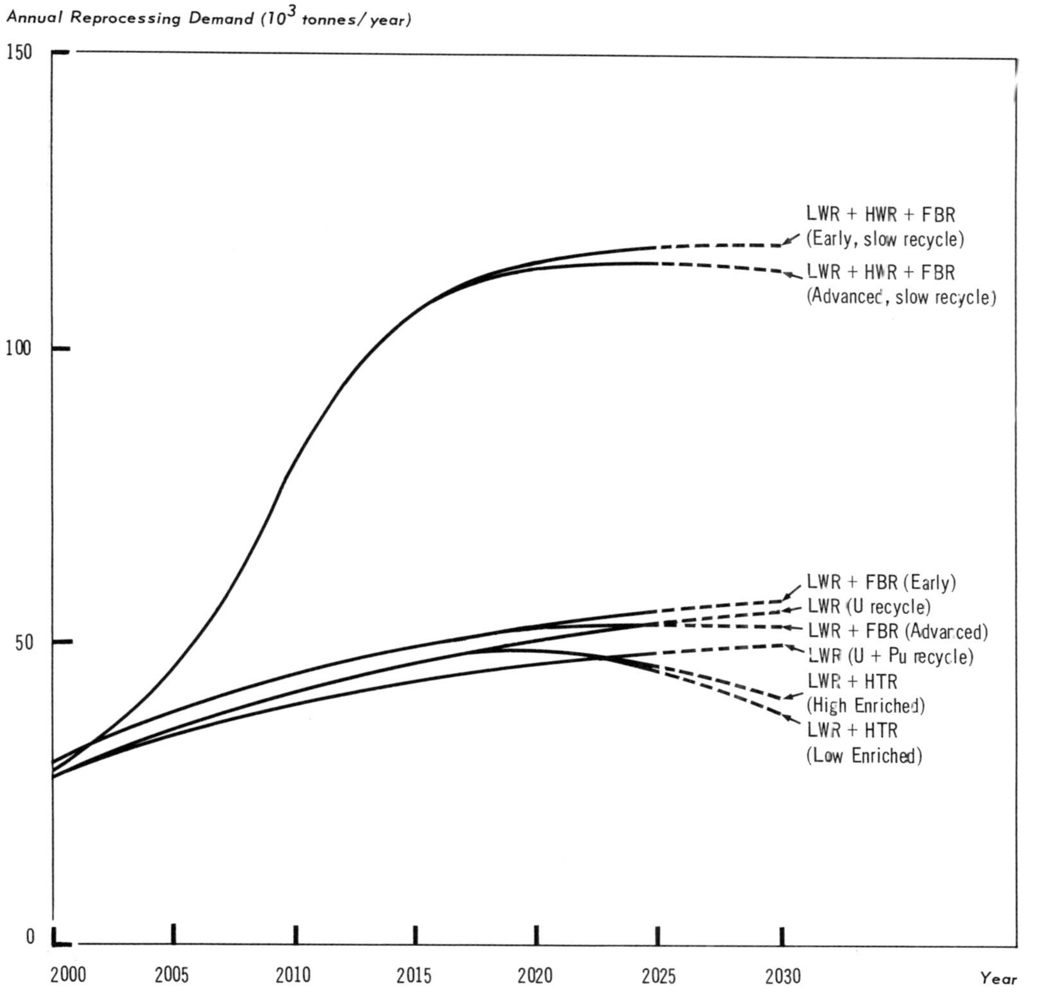

of course to be hoped that this resolution will take place in time to permit the emergence of a viable industry that must cope with its greater responsibilities in the longer-term.

F. FABRICATION

1. Demand

The demand for fabrication services is shown in Figures 27 and 28, for all reactor scenarios.

2. Supply

As noted in the short-term, the importance of nuclear fuel fabrication, an "upstream" part of the fuel cycle, was acknowledged at an early stage in the industry's development. However, aside from the order-of-magnitude increases in the amount of heavy metal to be fabricated in the longer-term, is the fact that the industry has not yet developed plant for fabricating on a large-scale, mixed-oxide LWR fuel. Further, although FBR fuel fabrication is in principle similar to that for LWR mixed-oxide fuel, in practice it is considerably more complex. Thus, the industry must prepare itself for the time when it must meet demands of the kind and magnitude associated with the longer-term.

Figure 27

WORLD ANNUAL FABRICATION DEMAND
2000-2025

Extension of Accelerated Nuclear Power Growth

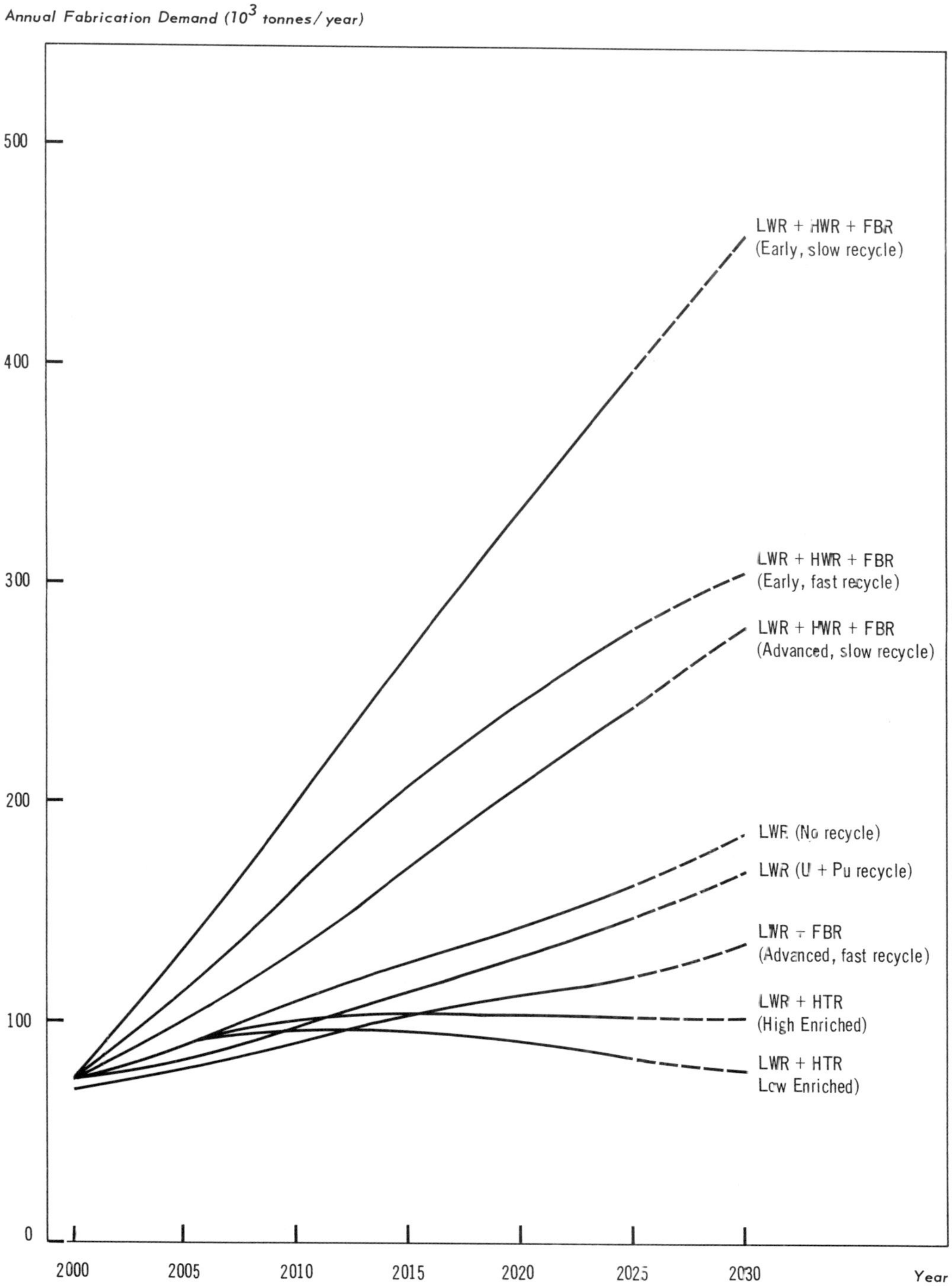

Figure 28

WORLD ANNUAL FABRICATION DEMAND
2000-2025

Extension of Present Trend Nuclear Power Growth

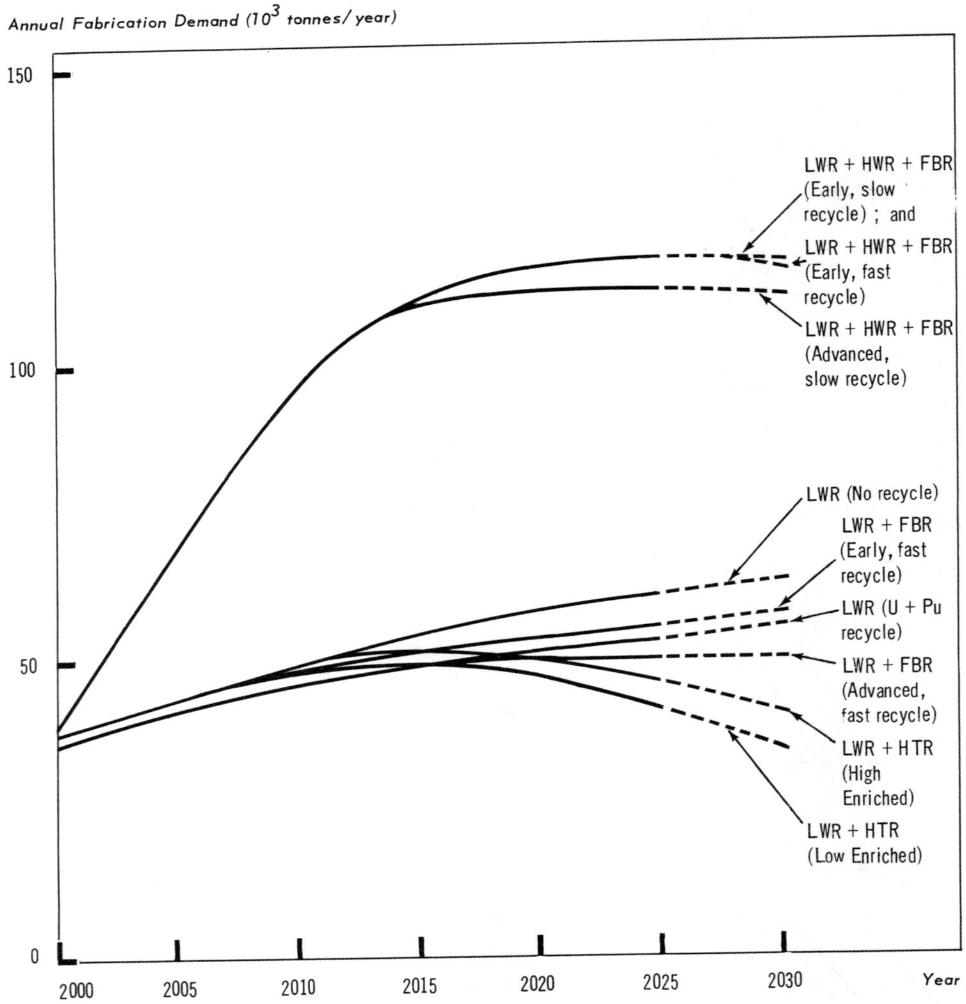

REFERENCES

/1/ "World Energy Outlook", OECD, Paris, 1977.

/2/ "Uranium Resources, Production and Demand", a joint report by the OECD Nuclear Energy Agency and the International Atomic Energy Agency, December 1977.

/3/ J. Miida, W. Häussermann, and S. Mankin, "Nuclear Power Programmes and Medium Term Projections in the OECD Area". International Conference on Nuclear Power and its Fuel Cycle, Salzburg, Austria, 2-13 May 1977, Paper IAEA-CN-36/492.

/4/ K.R. Williams, Shell International Petroleum Co., Ltd., "Projected Energy Requirements up to the year 2000", International Symposium on Uranium Supply and Demand, The Uranium Institute, London, 22-24 June 1977.

/5/ André Giraud, "World Energy Resources", Conference on World Nuclear Power, Atomic Industrial Forum, Washington, 15 November 1976.

/6/ R.D. Vaughan, C. Buck and T.N. Marsham, "The Practical Issues of the Fast Breeder as an Energy Source", Tenth World Energy Conference, Istanbul, September 1977.

/7/ "Energy: Global Prospects 1985-2000", Report of the Workshop on Alternative Energy Strategies (WAES), McGraw-Hill Book Company, 1977.

/8/ W. Häfele and W. Sassin, "Energy Strategies", Report RR-76-8, International Institute for Applied Systems Analysis, Laxenburg, Austria, March 1976.

/9/ J.C. Charrault, G. Graziani, J.H. de Nordwall, C. Rinaldini, R. Viseur, C. Vivante, and C. Zanantoni, "The European Community Case for Nuclear Power". International Conference on Nuclear Power and its Fuel Cycle, Salzburg, Austria, 2-13 May 1977, Paper IAEA-CN-36/365.

/10/ E.J. Hanrahan, R.H. Williamson, and R.W. Bown, "World Requirements and Supply of Uranium", International Conference on Uranium, Atomic Industrial Forum, Geneva, 14 September 1976.

/11/ John C. Fisher and Robert H. Pry, "A Simple Substitution Model of Technological Change", Chapter XVII, "Industrial Applications of Technological Forecasting", John Wiley and Sons, 1971.

/12/ United States Federal Energy Administration, Forecast for Nuclear Power, June 1977.

/13/ M.F. Duret, et. al., "The Contribution of Nuclear Power to World Energy Supply, 1975 to 2020", a Report prepared for the Conservation Commission of the World Energy Conference, Ottawa, Canada, July 1977.

/14/ J.A. Lane, A.J. Covarrubias, B.J. Csik, A. Fattah, and G. Woite, "Nuclear Power in Developing Countries". International Conference on Nuclear Power and its Fuel Cycle, Salzburg, Austria, 2-13 May 1977, Paper IAEA-CN-36/500.

/15/ Klaus Peter Messer, "Uranium Demand as Judged by Electric Utilities", International Symposium on Uranium Supply and Demand, The Uranium Institute, London, 22-24 June 1977.

/16/ U. Braatz and H.-J. Dibbert, "Kernbrennstoffversorgung: Sicherung durch Plafondpreise und engere Kooperation", Jahrbuch der Atomwirtschaft, 1977.

/17/ "FURE, A Fortran IV Programme for the Calculation of Fuel Cycle Requirements for a Long Term Nuclear Programme", N.V. tot Keuring van Elektrotechnische Materialen (N.V. KEMA), Arnhem, Netherlands.

/18/ J. Kostuik, "Industrial and Commercial Considerations Affecting Future Uranium Supply and Demand". International Conference on Nuclear Power and its Fuel Cycle, Salzburg, Austria, 2-13 May 1977, Paper IAEA-CN-36/55.

/19/ P.R. Vanstrum and S.A. Levin, "New Processes for Uranium Isotope Separation". International Conference on Nuclear Power and its Fuel Cycle, Salzburg, Austria, 2-13 May, 1977, Paper IAEA-CN-36/12.

/20/ J.F. Petit, "Uranium Enrichment by the Gaseous Diffusion Process". International Conference on Nuclear Power and its Fuel Cycle, Salzburg, Austria, 2-13 May 1977, Paper IAEA-CN-36/223.

/21/ D. Aston and E. Raetz, "Status of Urenco/Centec Centrifuge Project and Advantages of the Process". International Conference on Nuclear Power and its Fuel Cycle, Salzburg, Austria, 2-13 May 1977, Paper IAEA-CN-36/99.

/22/ A.J.A. Roux, W.L. Grant, R.A. Barbour, R.S. Loubser, J.J. Wannenburg, "Development and Progress on the South African Uranium Enrichment Project". International Conference on Nuclear Power and its Fuel Cycle, Salzburg, Austria, 2-13 May 1977, Paper IAEA-CN-36/300.

/23/ E. Svenke, "Availability of Enrichment Services". International Conference on Nuclear Power and its Fuel Cycle, Salzburg, Austria, 2-13 May 1977, Paper IAEA-CN-36/276.

/24/ "Atomic Industrial Forum Group Urges Washington to Take Four Steps on Fuel-Storage Outlook", Nuclear Fuel, June 13, 1977.

/25/ "Reprocessing of Spent Nuclear Fuels in OECD Countries", a report by an Expert Group of the OECD Nuclear Energy Agency, January 1977.

/26/ "L'industrie nucléaire française", Commissariat à l'énergie atomique, France, 1977.

/27/ G.K. Rossney, "Status and Prospects for Reprocessing". International Conference on Nuclear Power and its Fuel Cycle, Salzburg, Austria, 2-13 May 1977, Paper IAEA-CN-36/56.

/28/ J.J. Went, "Status and Prospects of Thermal Breeders". International Conference on Nuclear Power and its Fuel Cycle, Salzburg, Austria, 2-13 May 1977, Paper IAEA-CN-36/302.

/29/ H. Hunt, "The Problem of forecasting the adequacy of uranium supplies from the standpoint of a nuclear R and D organisation", IAEA Advisory Group Meeting of Experts on Evaluation of Uranium Resources, Rome, November 1976.

/30/ R.D. Nininger, Conference on Nuclear power and applications in Latin America, Mexico City, Sept./Oct. 1975.

/31/ W. Häussermann, P. Hogroian, R. Krymm, J. Cameron, O. Petersen, and G. Woite, "Demand and Supply Estimates for the Nuclear Fuel Cycle". International Conference on Nuclear Power and its Fuel Cycle, Salzburg, Austria, 2-13 May 1977, Paper IAEA-CN-36/493.

Appendix I

THE SHORT-TERM
(1977-2000)

Table I-1. TOTAL ENERGY REQUIREMENTS (EJ, 10^{18} joules) AND GROWTH RATES, %/Year

	ESTIMATE*	1975	GROWTH RATE 73-75	1980	GROWTH RATE 75-80	1985	GROWTH RATE 80-85	1990	GROWTH RATE 85-90	2000	GROWTH RATE 90-2000
OECD (European Region) ..	ACC.	47.55	- 1.8	60.70	5.0	75.64	4.5	92.01	4	129.8	3.5
	P.T.			57.85	4.0	68.73	3.5	79.70	3	97.12	2
OECD (North American Region)	ACC.	78.57	- 2.0	97.91	4.5	119.1	4	141.5	3.5	190.1	3
	P.T.			91.09	3.0	105.6	3.0	119.5	2.5	138.6	1.6
OECD (Pacific Region) ...	ACC.	16.70	- 0.2	25.74	9	36.13	7	48.31	6	78.70	5
	P.T.			22.40	6	28.59	5	34.79	4	46.76	3
OECD (Total)	ACC.	142.82	- 1.7	184.35	5.2	230.87	4.6	281.8	4.1	398.6	3.5
	P.T.			171.34	3.7	202.92	3.4	234.0	2.9	282.5	1.9
Non-OECD Countries	ACC.	26.37	2.3	37.84	7.5	54.33	7.5	78.03	7.5	160.8	7.5
	P.T.			33.61	5	42.95	5	54.84	5	89.29	5
World	ACC.	169.19	- 1.1	222.2	5.6	285.2	5.1	359.8	4.8	559.4	4.5
	P.T.			205.0	3.9	245.9	3.7	288.8	3.3	371.8	2.7

* "ACC." and "P.T." estimates correspond to "Accelerated" and "Present Trend" conditions for energy growth as described in the text.

Table I-2. TOTAL ELECTRICAL REQUIREMENTS AND CAPACITIES (Twh, GWe) AND NUCLEAR POWER GROWTH ESTIMATE (GWe) IN THE SHORT-TERM

	ESTIMATE**	1977	1980				1985				1990				1995				2000			
			ACC.		P.T.		ACC.		P.T.		ACC.		P.T.		ACC.		P.T.		ACC.		P.T.	
				*		*		*		*		*		*		*		*		*		*
Electricity Requirement (Twh)	OECD Europe	1,577	2,094	9.9	1,854	5.6	3,015	7.6	2,405	5.3	3,936	5.5	2,945	4.1	5,108	5.4	3,438	3.1	6,243	4.1	3,748	1.7
	OECD America	2,509	3,046	6.7	2,785	3.6	4,170	6.5	3,281	3.3	5,365	5.2	3,846	3.2	6,816	4.9	4,482	3.1	8,503	4.5	4,852	1.6
	OECD Pacific	555	776	11.8	653	5.6	1,188	8.9	862	5.7	1,690	7.3	1,079	4.6	2,370	7.0	1,326	4.2	3,213	6.3	1,591	3.7
	OECD Total	4,641	5,916	8.4	5,292	4.5	8,373	7.2	6,548	4.4	10,991	5.6	7,870	3.7	14,294	5.4	9,246	3.3	17,959	4.7	10,191	2.0
	Non-OECD	580	780	10.3	700	6.4	1,240	9.7	980	7.0	1,966	9.7	1,380	7.1	3,015	8.9	2,004	7.8	4,424	8.0	2,700	6.1
	World	5,221	6,696	8.8	5,992	4.7	9,613	7.5	7,528	4.7	12,957	6.1	9,250	4.2	17,309	6.0	11,250	4.0	22,383	5.3	12,891	2.8
Electrical Power Capacity (GWe)	OECD Europe	359	469	9.3	423	5.5	662	7.1	528	4.5	848	5.1	672	4.9	1,080	5.0	785	3.2	1,296	3.7	855	1.7
	OECD America	573	682	6.0	636	3.6	915	6.1	755	3.5	1,156	4.8	878	3.1	1,440	4.5	1,023	3.1	1,765	4.2	1,108	1.6
	OECD Pacific	128	174	10.8	149	5.5	260	8.4	189	4.9	382	8.0	246	5.4	541	7.2	304	4.3	668	4.3	367	3.8
	OECD Total	1,060	1,325	7.7	1,208	4.5	1,837	6.8	1,472	4.0	2,386	5.4	1,796	4.0	3,061	5.1	2,112	3.3	3,729	4.0	2,330	2.0
	Non-OECD	158	210	10.0	190	6.3	330	9.5	250	5.6	510	9.1	375	8.5	765	8.0	558	8.3	1,098	7.5	789	7.2
	World	1,218	1,535	8.0	1,398	4.7	2,167	7.1	1,722	4.3	2,896	5.8	2,171	4.7	3,826	5.8	2,670	4.2	4,827	4.8	3,119	3.2
Nuclear Power Capacity (GWe)	OECD Europe	26	60	23.9	60	23.9	146	20.7	107	13.0	273	13.3	195	12.6	418	8.9	260	6.5	560	6.0	310	3.6
	OECD America	51	66	9.0	66	9.0	157	18.9	125	13.6	287	12.8	214	11.5	514	12.4	310	7.6	810	9.5	400	5.2
	OECD Pacific	8	15	14.5	15	14.5	40	21.7	27	12.5	80	14.9	50	13.1	158	14.6	90	12.5	270	11.3	140	9.2
	OECD Total	85	141	14.9	141	14.9	343	20.0	259	13.2	640	13.3	459	12.1	1,090	11.2	660	7.7	1,640	8.5	850	5.2
	Non-OECD	2	5	35.9	5	35.9	25	38.0	19	32.0	60	19.1	45	17.6	130	16.7	90	15.8	250	14.0	150	12.3
	World	87	146	15.4	146	15.4	368	20.8	278	14.1	700	13.7	504	12.5	1,220	11.8	750	8.4	1,890	9.1	1,000	6.7
Nuclear Share of Electrical Capacity (%)	OECD Europe	7	13		14		22		20		32		29		39		33		43		36	
	OECD America	9	9		10		17		17		25		24		36		30		46		36	
	OECD Pacific	6	10		10		15		14		21		20		29		30		40		38	
	OECD Average	8	11		12		19		18		27		26		36		31		44		36	
	Non-OECD	1	2		3		8		8		12		12		17		16		23		19	
	World	7	9		10		17		16		24		23		32		28		39		32	

NOTES: * Annual growth rate in percent.
** "ACC." and "P.T." estimates correspond to "Accelerated" and "Present Trend" conditions for energy growth.

Table I-3. NUCLEAR POWER GROWTH ("PRESENT TREND"), GW(e)

END OF YEAR:	77	78	79	80	81	82	83	84	85	86	87	88	89	90	95	2000
OECD Europe	26	36	49	60	67	77	86	97	107	122	141	159	179	195	260	310
OECD North America	51	54	60	66	74	84	96	110	125	142	160	178	196	214	310	400
OECD Pacific	8	12	13	15	17	19	22	24	27	31	35	39	44	50	90	140
OECD Total	85	102	122	141	158	180	204	231	259	295	336	376	419	459	660	850
Non-OECD Countries	2	3	4	5	7	9	12	15	19	23	28	33	39	45	90	150
World Total	87	105	126	146	165	189	216	246	278	318	364	409	458	504	750	1,000

Table I-4. REACTORS IN OPERATION, AND REACTORS NOW ALREADY UNDER CONSTRUCTION THAT COULD BE IN OPERATION IN THE YEARS INDICATED, GW(e)

COUNTRY	END OF YEAR								
	77	78	79	80	81	82	83	84	85
Austria	-	0.7	0.7	0.7	0.7	0.7	0.7	0.7	0.7
Belgium	1.7	1.7	1.7	1.7	2.6	2.6	3.5	3.5	3.5
Denmark	-	-	-	-	-	-	-	-	-
Finland	0.4	0.4	1.5	1.5	1.5	2.2	2.2	2.2	2.2
France	4.7	6.5	12	14	16	18	18	21	23
Germany, F.R. of	6	9	10.3	12	14	16	18	21	21
Greece	-	-	-	-	-	-	-	-	-
Ireland	-	-	-	-	-	-	-	-	-
Italy	0.6	1.4	1.4	1.4	1.4	1.4	1.4	1.4	1.4
Luxembourg	-	-	-	-	-	-	-	-	-
Netherlands	0.5	0.5	0.5	0.5	0.5	0.5	0.5	0.5	0.5
Norway	-	-	-	-	-	-	-	-	-
Portugal	-	-	-	-	-	-	-	-	-
Spain	1.1	2.1	4.1	8	8	8	11	13	13
Sweden	3.2	4.7	5.6	6.5	6.5	7.4	7.4	7.4	7.4
Switzerland	1.0	1.0	1.0	1.9	1.9	1.9	1.9	2.8	2.8
Turkey	-	-	-	-	-	-	-	-	-
United Kingdom	6.6	6.6	10.3	10.3	10.3	10.3	10.3	10.3	10.3
Total, European Region	26	35	49	58	63	69	75	84	86
Canada	3.3	4	5	6	6	7	8	9	10
United States	48	50	55	60	68	77	88	99	113
Total, North American Region .	51	54	60	66	74	84	96	108	123
Australia	-	-	-	-	-	-	-	-	-
Japan	8	12	13	15	17	17	17	17	17
New Zealand	-	-	-	-	-	-	-	-	-
Total, Pacific Region	8	12	13	15	17	17	17	17	17
OECD Total	85	101	122	139	154	170	188	209	226
Non-OECD Countries	2	3	4	5	7	9	12	15	19
World Total	87	104	126	145	161	179	200	224	245

Table I-5. NUCLEAR POWER GROWTH ("ACCELERATED"), GWe

REGION	END OF YEAR															
	77	78	79	80	81	82	83	84	85	86	87	88	89	90	95	2000
European	26	36	49	60	68	84	102	121	146	168	193	220	244	273	418	560
North American	51	54	60	66	77	90	110	132	157	179	201	227	255	287	514	810
Pacific Region	8	12	13	15	21	25	30	34	40	47	55	63	71	80	158	270
OECD Total	85	102	122	141	166	199	242	287	343	394	449	510	570	640	1,090	1,640
Non-OECD Countries Total	2	3	4	5	9	12	16	21	25	31	38	45	52	60	130	250
World Total	87	105	126	146	175	211	258	308	368	425	487	555	622	700	1,220	1,890

Table I-6. OFFICIAL RESPONSES TO QUESTIONNAIRES ON NUCLEAR POWER GROWTH, GWe (MID-1976)*

COUNTRY	END OF YEAR														
	76	77	78	79	80	81	82	83	84	85	86	87	88	89	90
Belgium	1.7	1.7	1.7	2.6	3.5	4.5	5.5	6.8	8.1	8.1	9.4	10.7	12	13	14.6
Denmark	-	-	-	-	-	-	-	-	-	-	0.9	1.8	1.8	2.7	2.7
France	5.0	5.8	7.8	12	15	20	25	30	35	40	45	50	56	62	70
Germany, F.R. .	7	9.1	11	14	19	24	29	35	39	45	51	55	60	65	70
Ireland	-	-	-	-	-	-	-	-	0.7	0.7	0.7	1.4	1.4	1.4	2.0
Italy	1.4	1.4	1.4	1.4	1.4	1.4	1.4	3.4	7.4	11.4	15.4	20	25	30	35
Luxembourg ...	-	-	-	-	-	-	-	-	1.2	1.2	1.2	1.2	1.2	1.2	1.2
Netherlands	0.5	0.5	0.5	0.5	0.5	0.5	0.5	0.5	0.5	1.5	1.5	2.5	2.5	3.5	3.5
United Kingdom .	6.9	8.3	10	11	11	11	12	12.5	13	15.7	18	19	20	23	25
Austria	-	0.7	0.7	0.7	0.7	0.7	2	2	2	3	3	4	4	5	5
Finland	-	0.4	1.5	1.5	2.2	2.2	2.2	2.2	2.2	3.2	3.2	4.2	4.2	5.2	5.2
Greece	-	-	-	-	-	-	-	-	0.6	0.6	0.6	1.2	1.2	1.2	1.2
Portugal	-	-	-	-	-	-	-	-	-	1.3	1.3	2.3	2.3	3.2	3.2
Spain	1.1	2.5	4.3	6.3	8.2	11.2	14.2	17.2	20.2	23.2	25	27	29	31	33
Sweden	3.2	4.2	4.8	5.4	6.0	6.7	7.4	8.0	9.0	10.0	11	11	12	13	13
Switzerland	1	1	1	1.9	1.9	2.5	2.5	3.3	3.8	3.8	3.8	3.8	3.8	3.8	4.2
Turkey	-	-	-	-	-	-	-	-	-	0.6	0.6	1.6	1.6	2.2	2.2
European Total .	28	36	45	57	69	85	102	121	143	169	192	217	238	266	291
Canada	3.3	4.0	5.0	6.0	7.0	8.0	8.8	9.6	11	12	13	15	18	21	25
United States ...	47	54	61	70	81	90	105	116	135	160	175	200	225	250	284
North American Total	50	58	66	76	88	98	114	126	146	172	188	215	243	271	309
Australia	-	-	-	-	-	-	-	-	-	-	-	-	0.5	0.5	1.0
Japan	9	11	13	15	17	24	30	36	41	49	54	60	67	73	83
New Zealand	-	-	-	-	-	-	-	-	-	-	-	-	1.2	1.2	1.2
Pacific Region Total	9	11	13	15	17	24	30	36	41	49	54	60	69	75	85
OECD Total	87	105	124	148	174	207	246	283	330	390	434	492	550	612	685
Non-OECD Countries**.....	1	2	3	4	5	9	12	16	21	25	31	38	45	52	60
World Total	88	107	127	152	179	216	258	299	351	415	465	530	595	664	748

* It should be noted that in several instances, official estimates have, since this date, been reduced to levels below those indicated in this Table. For example, for the years 1985 and 1990, the Federal Republic of Germany plans to have 30 GW(e) and 47 GW(e) installed, while the official figures for the United States are 127 GW(e) and 195-210 GW(e), respectively.

** Questionnaires were not sent to Non-OECD Countries, The figures shown are "Accelerated" power growth estimates.

Table I-7. WORLD REACTOR DISTRIBUTION
(1977-2000), GWe

YEAR	LWR		HWR		AGR		GG		HTR		FBR		TOTAL WORLD	
	ACCEL-ERATED	PRESENT TREND	ACCEL-ERATED	PRESENT TREND	ACCEL-ERATED	PRESENT TREND	ACCEL-ERATED	PRESENT TREND	ACCEL-ERATED	PRESENT TREND	ACCEL-ERATED	PRESENT TREND	ACCEL-ERATED	PRESENT TREND
1977	74	74	4	4	3	3	6	6	-	-	-	-	87	87
1978	90	90	5	5	3	3	6	6	-	-	1	1	105	105
1979	107	107	6	6	6	6	6	6	-	-	1	1	126	126
1980	126	126	7	7	6	6	6	6	-	-	1	1	146	146
1981	152	145	9	7	6	6	6	6	1	-	1	1	175	165
1982	185	167	12	9	6	6	6	6	1	-	1	1	211	189
1983	229	191	14	11	6	6	6	6	1	1	2	1	258	216
1984	278	220	15	12	6	6	6	6	1	1	2	1	308	246
1985	334	251	19	13	6	6	6	6	1	1	2	1	368	278
1986	387	288	22	16	6	6	6	6	2	1	2	1	425	318
1987	444	331	26	19	6	6	6	6	2	1	3	1	487	364
1988	507	372	30	22	6	6	6	6	2	1	4	2	555	409
1989	569	418	34	25	6	6	6	6	2	1	5	2	622	458
1990	641	460	39	28	6	6	6	6	2	2	6	2	700	504
1991	723	502	46	32	6	6	6	6	2	2	7	2	790	550
1992	814	548	53	36	6	6	5	5	3	2	9	3	890	600
1993	914	594	61	40	6	6	5	5	3	2	11	3	1,000	650
1994	1,012	637	71	45	6	6	5	5	3	3	13	4	1,110	700
1995	1,110	683	81	50	6	6	4	4	3	3	16	4	1,220	750
1996	1,232	727	93	55	6	6	4	4	4	3	21	5	1,360	800
1997	1,340	768	104	60	6	6	3	3	5	3	42	5	1,500	845
1998	1,423	815	117	65	6	6	3	3	6	4	75	7	1,630	900
1999	1,501	859	129	70	6	6	2	2	7	5	105	8	1,750	950
2000	1,585	901	142	75	6	6	2	2	10	6	145	10	1,890	1,000

NOTES:

Reactor abbreviations: LWR-Light Water Reactor; HWR-Heavy Water Reactor; AGR-Advanced Gas Reactor; GG-Graphite Gas Reactor; HTR-High Temperature Gas Reactor; FBR-Fact Breeder Reactor.

Reactor lifetime: 30 years.

Table I-8. OECD REACTOR DISTRIBUTION IN THE SHORT-TERM, GWe *

YEAR	LWR		HWR		AGR		GG		HTR		FBR		TOTAL (OECD)	
	ACC.	P.T.	ACC.	P.T.	ACC.	P.T.	ACC.	P.T.	ACC.	P.T.	ACC.	P.T.	ACC.	P.T.
1977 ..	72	72	4	4	3	3	6	6	-	-	-	-	85	85
1978 ..	88	88	4	4	3	3	6	6	-	-	1	1	102	102
1979 ..	104	104	5	5	6	6	6	6	-	-	1	1	122	122
1980 ..	122	122	6	6	6	6	6	6	-	-	1	1	141	141
1981 ..	146	139	6	6	6	6	6	6	1	-	1	1	166	158
1982 ..	178	160	7	7	6	6	6	6	1	-	1	1	199	180
1983 ..	218	182	9	8	6	6	6	6	1	1	2	1	242	204
1984 ..	262	208	10	9	6	6	6	6	1	1	2	1	287	231
1985 ..	315	235	13	10	6	6	6	6	1	1	2	1	343	259
1986 ..	363	269	15	12	6	6	6	6	2	1	2	1	394	295
1987 ..	414	308	18	14	6	6	6	6	2	1	3	1	449	336
1988 ..	472	345	20	16	6	6	6	6	2	1	4	2	510	376
1989 ..	528	386	23	18	6	6	6	6	2	1	5	2	570	419
1990 ..	593	423	27	20	6	6	6	6	2	2	6	2	640	459
1991 ..	668	458	31	22	6	6	6	6	2	2	7	2	720	496
1992 ..	749	497	36	25	6	6	5	5	3	2	9	3	808	538
1993 ..	837	536	41	27	6	6	5	5	3	2	11	3	903	579
1994 ..	921	573	48	30	6	6	5	5	3	3	13	4	996	621
1995 ..	1,007	610	54	33	6	6	4	4	3	3	16	4	1,090	660
1996 ..	1,113	647	62	35	6	6	4	4	4	3	21	5	1,210	700
1997 ..	1,205	680	69	38	6	6	3	3	5	3	42	5	1,330	735
1998 ..	1,265	716	77	41	6	6	3	3	6	4	75	7	1,432	777
1999 ..	1,322	750	84	43	6	6	2	2	7	5	105	8	1,526	814
2000 ..	1,385	780	92	46	6	6	2	2	10	6	145	10	1,640	850

Table I-9. NON-OECD COUNTRIES' REACTOR DISTRIBUTION IN THE SHORT-TERM, GWe *

YEAR	LWR		HWR		TOTAL (NON-OECD COUNTRIES)	
	ACC.	P.T.	ACC.	P.T.	ACC.	P.T.
1977	2	2	0	0	2	2
1978	2	2	1	1	3	3
1979	3	3	1	1	4	4
1980	4	4	1	1	5	5
1981	6	6	3	1	9	7
1982	7	7	5	2	12	9
1983	11	9	5	3	16	12
1984	16	12	5	3	21	15
1985	19	16	6	3	25	19
1986	24	19	7	4	31	32
1987	30	23	8	5	38	28
1988	35	27	10	6	45	33
1989	41	32	11	7	52	39
1990	48	37	12	8	60	45
1991	55	44	15	10	70	54
1992	65	51	17	11	82	62
1993	77	58	20	13	97	71
1994	91	64	23	15	114	79
1995	103	73	27	17	130	90
1996	119	80	31	20	150	100
1997	135	88	35	22	170	110
1998	158	99	40	24	198	123
1999	179	109	45	27	224	136
2000	200	121	50	29	250	150

* "ACC." and "P.T." estimates correspond to "Accelerated" and "Present Trend" conditions for energy growth.

Figure I-1

PARAMETERS USED IN DERIVING NUCLEAR POWER FORECAST FROM TOTAL ENERGY REQUIREMENTS*

(for both the short and long-term)

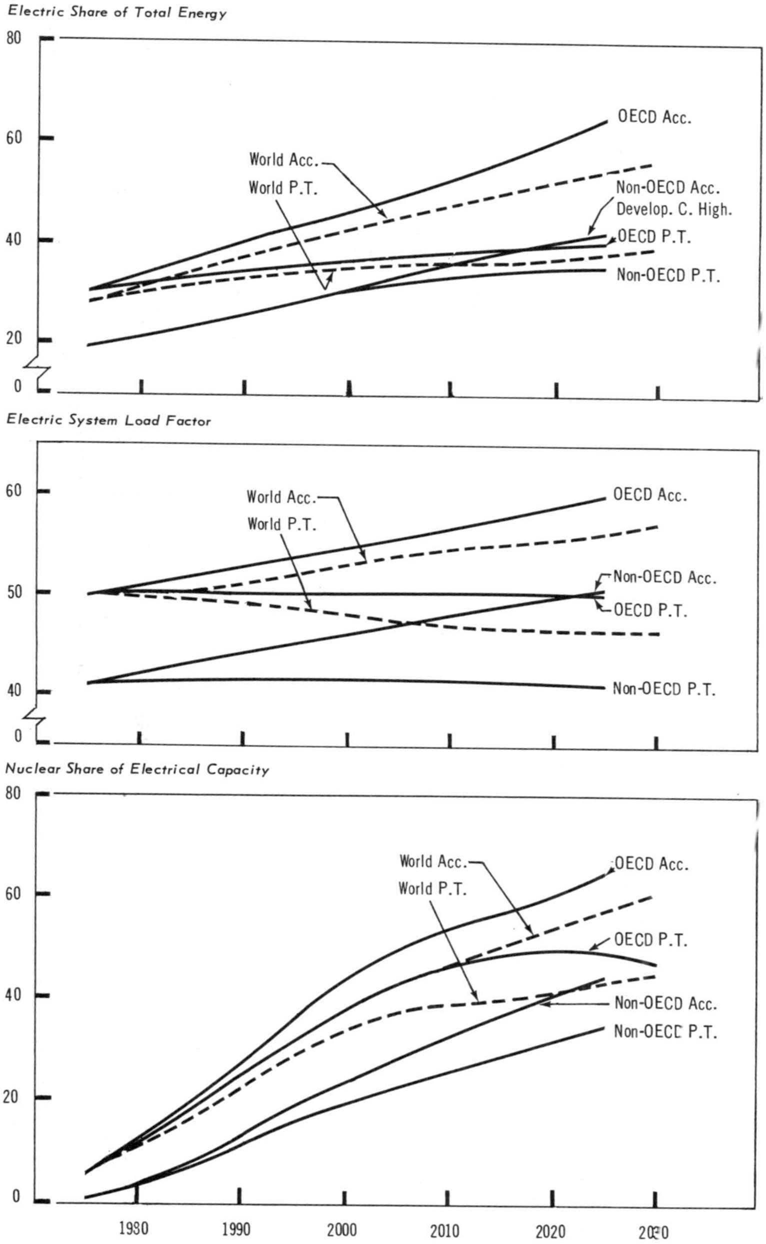

* Acc. = Accelerated Growth.
P.T. = Present Trend Growth.

Table I-10. POWER REACTOR FUEL CYCLE CHARACTERISTICS[(1)]
(Reactors marked by asterisk (*) used in short-term, 1977-2000)

REACTOR TYPE[(2)]	LWR *		HWR		HTR		GG*	AGR*	FBR	
	U CYCLE	Pu CYCLE	U CYCLE*	Th CYCLE	HIGH- * ENRICHMENT	LOW-ENRICHMENT			EARLY[(7)]*	ADVANCED
Conversion Ratio	0.55	0.53	0.71	1.00	0.72	0.55	0.86	0.44	1.16	1.43
Doubling Time, years[(8)]	-	-	-	-	-	-	-	-	30/60	12/24
Initial Loading										
Uranium (t/GWe)	90.7	77.4	143	3.3	1.27	19.97	900	155	56.4	68.1
Average initial enrichment (wt % U-235)	2.23	0.71	0.71	93	93	4.57	0.71	1.73	depleted	depleted
Natural Uranium required (t/GWe)	393	78.9	145	664	256	187	918	498	-	-
Separative work required (t SW/GWe)	213	-	-	711	274	140	-	223	-	-
Fissile plutonium (t Pu/GWe)	-	1.50	-	-	-	-	-	-	3.00	2.37
Thorium (t/GWe)	-	-	-	127	33.9	-	-	-	-	-
Replacement Loading										
Uranium (t/GWe.a)	35.7	25.8	168	1.65	1.23	11.27	300	46	26.0	34.2
Average enrichment (wt % U-235)	3.02	0.71	0.71	(10)	(4)	8.98	0.71	2.29	depleted	depleted
Natural uranium required (t/GWe.a)	218	26.4	170	(10)	88(4)	217	306	204	32.1	-
Separative work required (t SW/GWe.a)	140	-	-	(10)	95(4)	191	-	113	-	-
Fissile plutonium (t Pu/GWe.a)	-	1.27	-	-	-	-	-	-	2.00	1.89
Thorium (t/GWe.a)	-	-	-	97.5	15.2	-	-	-	-	-
Discharged Fuel										
Burn-up (GWd/t)	30.6	30.0	7.5	10.0	71	94	4.5	18.0	60	56
Uranium (t/GWe.a)	34.7	25.1	166	1.65	0.89	9.74		45	28.9	33.5
Average enrichment (wt % U-235)	0.85	0.20	0.20	(10)	(5)	1.53	0.4	0.76	depleted	depleted
(3) Natural uranium equivalent (t/GWe.a)	45	-	-	(10)	(6)	27	-	-	-	-
(3) Separative work equivalent (t SW/GWe.a)	4.9	-	-	(10)	(6)	11	-	-	-	-
Fissile plutonium (t Pu/GWe.a)[(9)]	0.22	0.86	0.50	-	-	0.11	0.60	0.22	2.21	2.28
Thorium (t/GWe.a)	-	-	-	96.5	14.3	-	-	-	-	-
Delays (short/long), years										
Mill out to reactor in — U fuel	1.83	1.83	1.33	1.83	1.83	1.83	1.33	1.83	-	-
Pu fuel	-	0.83	-	-	-	-	-	-	0.25/0.83	0.25/0.83
Reactor out to reactor in — U fuel	1.92/3.33	3.33	1.42/2.83	3.33	3.33	3.33	2.83	2.33	-	-
Pu fuel	-	2.83	-	-	2.83	2.83	2.83	2.83	0.75/2.83	0.75/2.83
Reprocessing out to reactor in — U fuel	1.33	1.22	1.33	1.33	1.33	1.33	1.33	1.33	-	-
Pu fuel	0.83	0.83	0.83	-	0.83	0.83	0.83	0.83	0.25/0.83	0.25/0.83
Enrichment out to reactor in	0.83	0.83	0.83	0.83	0.83	0.83	0.83	0.83	-	-
Fabrication out to reactor in	0.42	0.42	0.42	0.42	0.42	0.42	0.42	0.42	0.08/0.42	0.08/0.42

(1) Data given on basis of 100% load factor (i.e., reactor on-stream, at full capacity, for 8,760 hours/year). However, fuel cycle requirements presented in this report have been computed on basis of 70% load factor. Uranium and separative work requirements based on 0.25% U-235 enrichment plant tails assay.

(2) LWR = light water reactor (2:1 ratio of PWR:BWR); HWR = heavy water reactor; HTR = high-temperature gas-cooled reactor; GG = graphite-gas reactor; AGR = advanced gas reactor; FBR = fast breeder reactor.

(3) Savings, where stated, are due only to the recycling of uranium.

(4) Replacement fuel consists of 35.6 wt% fresh make-up fuel (93% enriched) and 64.4 wt% recycle fuel (assay of 47.4 wt% U-233 and 11.8 wt% U-235). Uranium and separative work equivalents shown are for fresh fuel only.

(5) Average assay is 43.3 wt% U-233 and 11.6 wt% U-235.

(6) Mixed isotopes; hence, computation of natural uranium and separative work savings must be correlated with reactivity of fresh fuel.

(7) Used as FBR model prior to year 2000, and as an alternative to the advanced FBR after year 2000.

(8) For short delays/long delays. Takes account of out-of-core inventories and delays, but is computed on basis of 100% utilization of all Pu produced and 100% load factor.

(9) 98% utilization assumed.

(10) Recycle fuel composition is 91 wt% U-233 and 8 wt% U-235. No additional uranium or separative work required if discharged fuel is recycled.

* Reactors utilized in the short-term.

Appendix II

THE LONGER-TERM
(2000-2025)

Table II-1. TOTAL ENERGY REQUIREMENTS, EJ (10^{18} joules) AND GROWTH RATES (%/year)

	*	2000	GROWTH RATE 1990-2000	2025	GROWTH RATE 1990-2000
OECD Europe	ACC.	129.8	3.5	194.1	1.6
	P.T.	97.12	2	97.12	0
OECD America	ACC.	190.1	3	251.8	1.1
	P.T.	138.6	1.5	138.6	0
OECD Pacific	ACC.	78.70	5	149.9	2.6
	P.T.	46.76	3	76.31	2
OECD Total	ACC.	399.8	3.5	595.8	1.6
	P.T.	282.5	1.9	312.5	0.4
Non-OECD Countries	ACC.	160.8	7.5	492.6	4.6
	P.T.	89.29	5	238.0	4
World	ACC.	559.4	4.5	1,088.3	2.7
	P.T.	371.8	2.7	550.5	1.6

* "ACC" and "P.T." refer to extensions into the longer-term of "Accelerated" and "Present Trend" power growth estimates.

Table II-2. ELECTRICAL REQUIREMENTS AND CAPACITIES (TWh, GWe) NUCLEAR POWER GROWTH ESTIMATE (GWe) IN THE LONG-TERM**

		2005				2010				2015				2020				2025			
		ACC.		P.T.		ACC.		P.T.		ACC.		P.T.		ACC.		P.T.		ACC.		P.T.	
			*		*		*		*		*		*		*		*		*		*
Electricity Requirement (TWh)	OECD Total ..	22,227	4.4	11,151	1.8	26,474	3.6	11,427	0.5	30,658	3.0	11,740	0.5	34,990	2.7	11,979	0.4	38,148	1.7	12,150	0.3
	Non-OECD Countries	6,391	7.6	3,709	6.6	9,070	7.3	5,096	6.6	12,761	5.1	6,214	4.0	16,026	4.7	7,358	3.4	19,854	4.4	8,099	1.9
	World	28,618	5.0	14,860	2.0	35,544	4.4	16,523	2.1	43,419	3.9	17,895	1.6	51,016	3.3	19,337	1.6	58,002	2.6	20,251	0.9
Electrical Capacity (GWe)	OECD Total ..	4,531	4.0	2,546	1.8	5,302	3.2	2,609	0.5	6,034	2.6	2,680	0.5	6,770	2.3	2,735	0.4	7,154	1.1	2,774	0.3
	Non-OECD Countries	1,569	7.4	1,052	5.9	2,180	6.8	1,385	5.7	2,973	6.4	1,689	4.0	3,659	4.2	2,000	3.4	4,444	4.0	2,201	1.9
	World	6,100	4.8	3,598	2.9	7,482	4.2	3,994	2.1	9,007	3.5	4,354	1.7	10,429	3.0	4,735	1.7	11,750	2.4	4,975	1.0
Nuclear power Capacity (GWe)	OECD Total ..	2,220	6.2	1,060	4.5	2,810	4.8	1,200	2.5	3,500	4.5	1,280	1.3	4,130	3.4	1,340	0.9	4,650	2.4	1,387	0.7
	Non-OECD Countries	460	13.0	242	10.0	780	11.1	360	8.3	1,100	7.1	490	6.4	1,500	6.4	640	5.5	2,000	5.9	770	3.8
	World	2,680	17.2	1,302	5.4	3,590	6.0	1,560	3.7	4,600	5.1	1,770	2.6	5,630	4.1	1,980	2.3	6,650	3.4	2,157	1.7

* Annual growth rate in percent (averaged over each preceeding 5-year time increment).

** ACC = Accelerated; P.T. = Present Trend.

Figure II-1

REACTOR DISTRIBUTION IN LONG TERM (2000-2025)

LWR + FBR Case

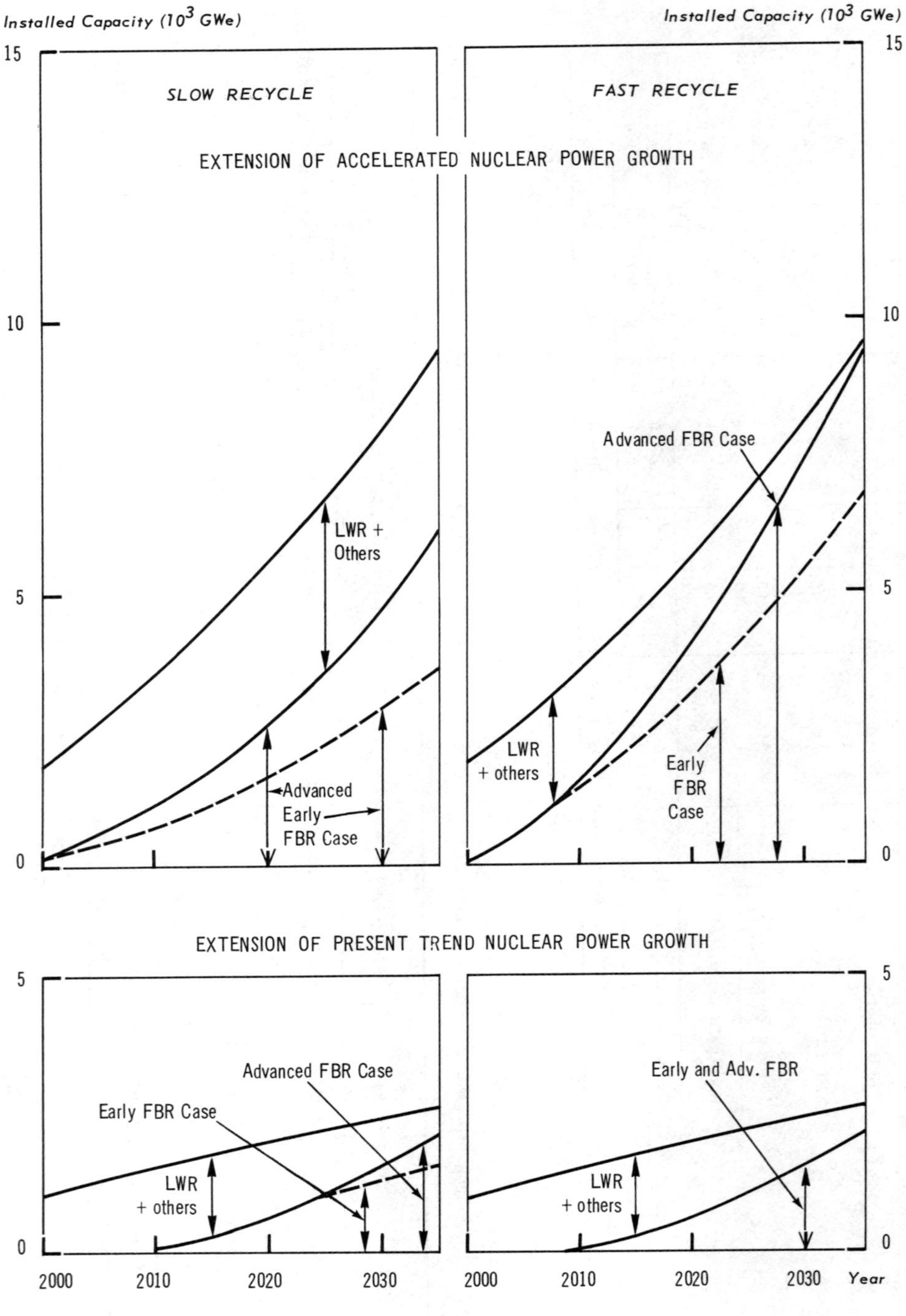

Figure II-2

REACTOR DISTRIBUTION IN LONG TERM (2000-2025)

LWR + HWR + Early FBR Case

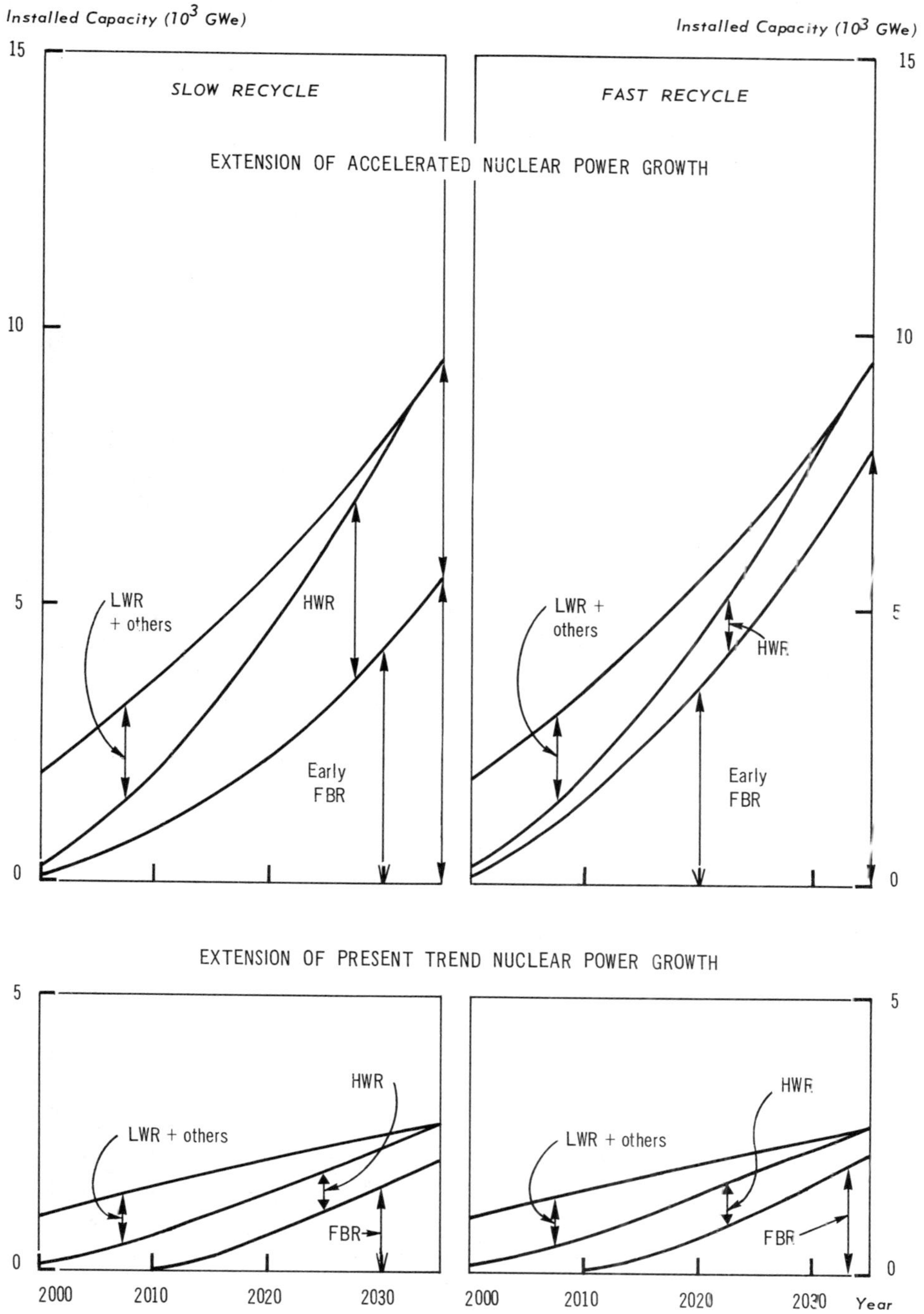

Figure II-3

REACTOR DISTRIBUTION IN LONG TERM (2000-2025)

LWR + HWR + Adv. FBR Case

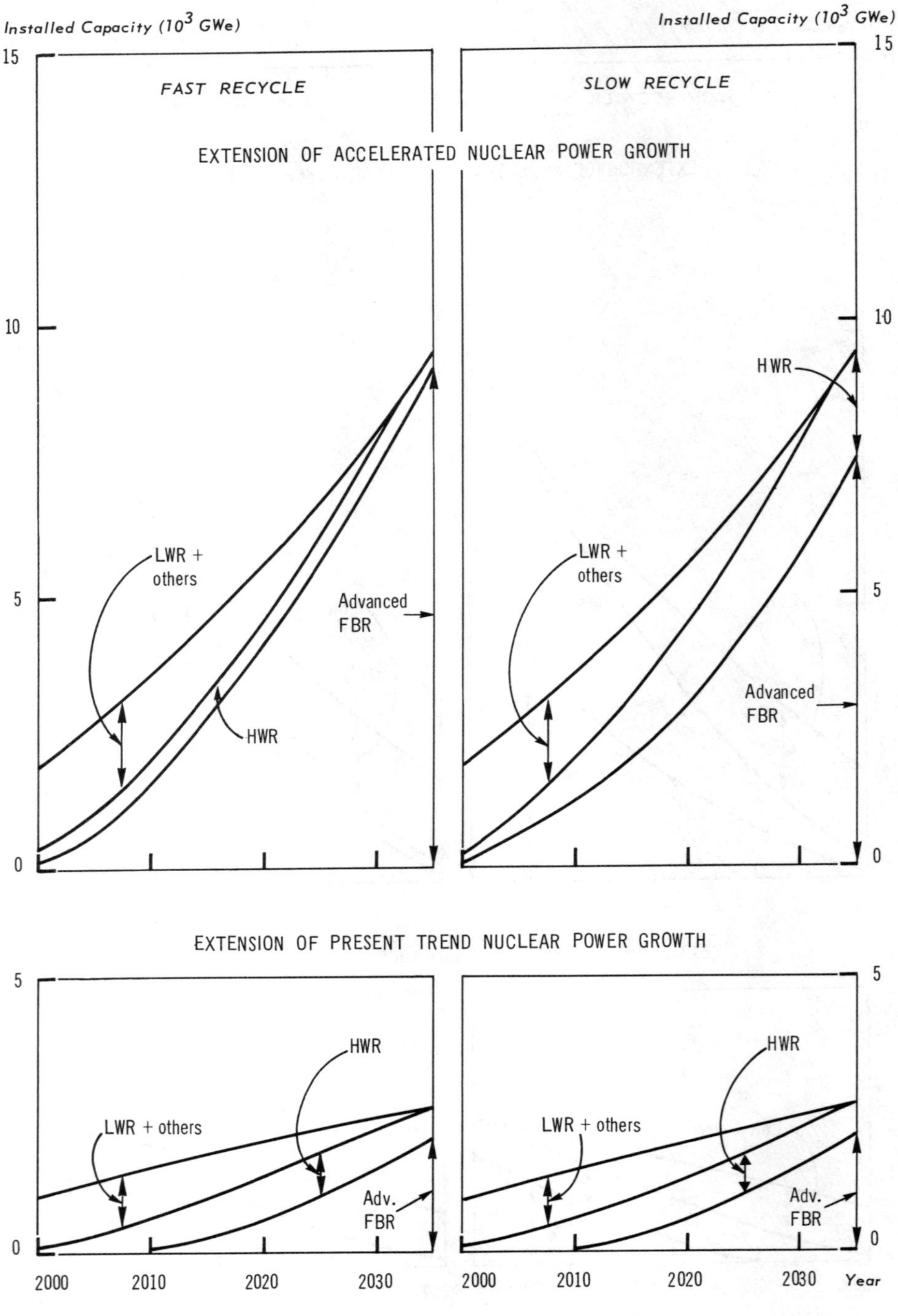

Figure II-4

REACTOR DISTRIBUTION IN LONG TERM (2000-2025)

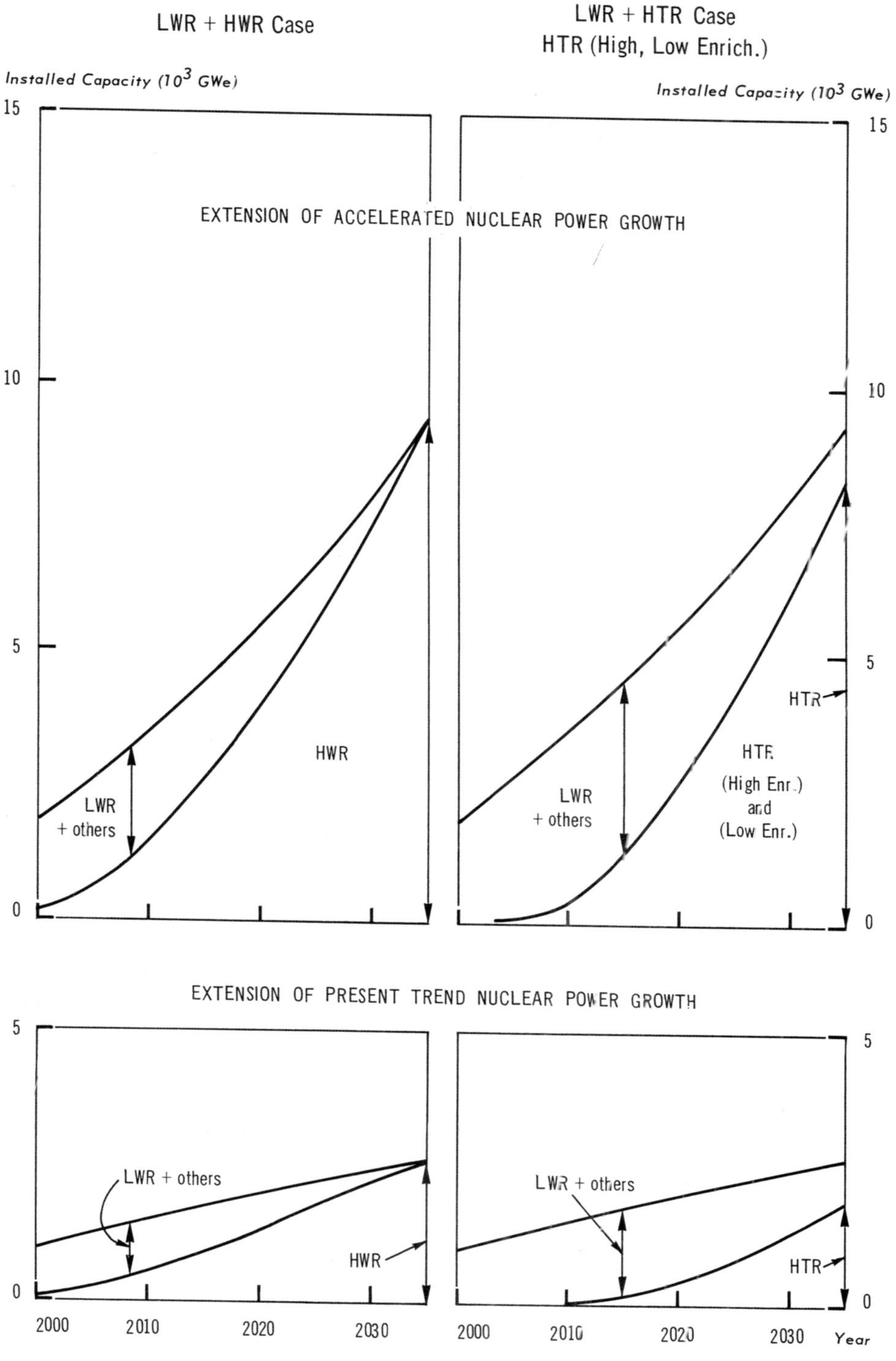

Appendix III

MEMBERS OF THE WORKING PARTY ON URANIUM DEMAND

Country		Name	Organisation
Belgium		P. Duflou	Synatom, Brussels
Canada		H.B. Merlin	Department of Energy, Mines and Resources, Ottawa
Finland		I. Mäkipentti	Ministry of Trade and Industry, Helsinki
France		L. Thiriet A. Ferrari	Commissariat à l'Energie Atomique, Paris "
Germany, Federal Republic of		U. Braatz	Vereinigung Deutscher Elektrizitätswerke, Frankfurt/Main
		G. Hauerstein	Bundesministerium für Forschung und Technologie, Bonn
Italy		P. Valant	Comitato Nazionale per l'Energia Nucleare, Rome
		M. Cuzzaniti V. Formisano	Ente Nazionale Idrocarburi, Rome "
		S. Valenziani	Ente Nazionale per l'Energia Elettrica, Rome
Japan		H. Satake T. Bito	Permanent Delegation of Japan to the OECD, Paris
Netherlands	Chairman:	W.K. Wiechers, M. J. Crijns	N.V. Kema, Arnhem
Portugal		M.A. Brandao	Permanent Delegation of Portugal to the OECD
Spain		J.M. Clement	Junta de Energia Nuclear, Madrid
Sweden		K. Jirlow P.H. Margen G. Vieider	A.B. Atomenergi, Nyköping " "
Switzerland		B. Hausherr	Federal Office of Energy, Berne
United Kingdom		A. Green	United Kingdom Atomic Energy Authority, Risley
		D.S. Briggs D.F.I. Bishop	British Nuclear Fuels, Ltd., Risley
CEC		J.C. Charrault A. Decressin M. van Meerbeeck	
IAEA Division of Nuclear Power and Reactors		R. Krymm G. Woite,	

SECRETARIAT

<u>OECD/NEA</u> Nuclear Development Division	W. Häussermann P. Hogroian S. Mankin

OECD SALES AGENTS
DÉPOSITAIRES DES PUBLICATIONS DE L'OCDE

ARGENTINA – ARGENTINE
Carlos Hirsch S.R.L., Florida 165,
BUENOS-AIRES, ☎ 33-1787-2391 Y 30-7122

AUSTRALIA – AUSTRALIE
International B.C.N. Library Suppliers Pty Ltd.,
161 Sturt St., South MELBOURNE, Vic. 3205. ☎ 699-6388
658 Pittwater Road, BROOKVALE NSW 2100. ☎ 938 2267

AUSTRIA – AUTRICHE
Gerold and Co., Graben 31, WIEN 1. ☎ 52.22.35

BELGIUM – BELGIQUE
Librairie des Sciences,
Coudenberg 76-78, B 1000 BRUXELLES 1. ☎ 512-05-60

BRAZIL – BRÉSIL
Mestre Jou S.A., Rua Guaipá 518,
Caixa Postal 24090, 05089 SAO PAULO 10. ☎ 261-1920
Rua Senador Dantas 19 s/205-6. RIO DE JANEIRO GB.
☎ 232-07. 32

CANADA
Renouf Publishing Company Limited,
2182 St. Catherine Street West,
MONTREAL, Quebec H3H 1M7 ☎ (514) 937-3519

DENMARK – DANEMARK
Munksgaards Boghandel,
Nørregade 6, 1165 KØBENHAVN K. ☎ (01) 12 69 70

FINLAND – FINLANDE
Akateeminen Kirjakauppa
Keskuskatu 1, 00100 HELSINKI 10. ☎ 625.901

FRANCE
Bureau des Publications de l'OCDE,
2 rue André-Pascal, 75775 PARIS CEDEX 16.
☎ 524.81.67
Principal correspondant :
13602 AIX-EN-PROVENCE : Librairie de l'Université.
☎ 26.18.08

GERMANY – ALLEMAGNE
Verlag Weltarchiv G.m.b.H.
D 2000 HAMBURG 36, Neuer Jungfernstieg 21.
☎ 040-35-62-500

GREECE – GRÈCE
Librairie Kauffmann, 28 rue du Stade,
ATHÈNES 132. ☎ 322.21.60

HONG-KONG
Government Information Services,
Sales and Publications Office, Beaconsfield House, 1st floor,
Queen's Road, Central. ☎ H-233191

ICELAND – ISLANDE
Snaebjörn Jónsson and Co., h.f.,
Hafnarstraeti 4 and 9, P.O.B. 1131, REYKJAVIC.
☎ 13133/14281/11936

INDIA – INDE
Oxford Book and Stationery Co.:
NEW DELHI, Scindia House. ☎ 45896
CALCUTTA, 17 Park Street. ☎ 240832

IRELAND - IRLANDE
Eason and Son, 40 Lower O'Connell Street,
P.O.B. 42, DUBLIN 1. ☎ 74 39 35

ISRAËL
Emanuel Brown: 35 Allenby Road, TEL AVIV. ☎ 51049/54082
also at:
9, Shlomzion Hamalka Street, JERUSALEM. ☎ 234807
48 Nahlath Benjamin Street, TEL AVIV. ☎ 53276

ITALY – ITALIE
Libreria Commissionaria Sansoni:
Via Lamarmora 45, 50121 FIRENZE. ☎ 579751
Via Bartolini 29, 20155 MILANO. ☎ 365083
Sous-dépositaires :
Editrice e Libreria Herder,
Piazza Montecitorio 120, 00 186 ROMA. ☎ 674628
Libreria Hoepli, Via Hoepli 5, 20121 MILANO. ☎ 365446
Libreria Lattes, Via Garibaldi 3, 10122 TORINO. ☎ 519274
La diffusione delle edizioni OCDE è inoltre assicurata dalle migliori

JAPAN – JAPON
OECD Publications Centre,
Akasaka Park Building, 2-3-4 Akasaka, Minato-ku,
TOKYO 107. ☎ 586-2016

KOREA - CORÉE
Pan Korea Book Corporation,
P.O.Box n°101 Kwangwhamun, SÉOUL. ☎ 72-7369

LEBANON – LIBAN
Documenta Scientifica/Redico,
Edison Building, Bliss Street, P.O.Box 5641, BEIRUT.
☎ 354429–344425

THE NETHERLANDS – PAYS-BAS
Staatsuitgeverij
Chr. Plantijnstraat
'S-GRAVENHAGE. ☎ 070-814511
Voor bestellingen: ☎ 070-624551

NEW ZEALAND - NOUVELLE-ZÉLANDE
The Publications Manager,
Government Printing Office,
WELLINGTON: Mulgrave Street (Private Bag),
World Trade Centre, Cubacade, Cuba Street,
Rutherford House, Lambton Quay, ☎ 737-320
AUCKLAND: Rutland Street (P.O.Box 5344), ☎ 32.919
CHRISTCHURCH: 130 Oxford Tce (Private Bag), ☎ 50.331
HAMILTON: Barton Street (P.O.Box 857), ☎ 80.103
DUNEDIN: T & G Building, Princes Street (P.O.Box 1104),
☎ 78.294

NORWAY – NORVÈGE
Johan Grundt Tanums Bokhandel,
Karl Johansgate 41/43, OSLO 1. ☎ 02-332980

PAKISTAN
Mirza Book Agency, 65 Shahrah Quaid-E-Azam, LAHORE 3.
☎ 66839

PHILIPPINES
R.M. Garcia Publishing House, 903 Quezon Blvd. Ext.,
QUEZON CITY, P.O.Box 1860 – MANILA. ☎ 99.98.47

PORTUGAL
Livraria Portugal, Rua do Carmo 70-74, LISBOA 2. ☎ 360582/3

SPAIN – ESPAGNE
Mundi-Prensa Libros, S.A.
Castelló 37, Apartado 1223, MADRID-1. ☎ 275.46.55
Libreria Bastinos, Pelayo, 52, BARCELONA 1. ☎ 222.06.00

SWEDEN – SUÈDE
AB CE FRITZES KUNGL HOVBOKHANDEL,
Box 16 356, S 103 27 STH, Regeringsgatan 12,
DS STOCKHOLM. ☎ 08/23 89 00

SWITZERLAND – SUISSE
Librairie Payot, 6 rue Grenus, 1211 GENÈVE 11. ☎ 022-31.89.50

TAIWAN – FORMOSE
National Book Company,
84-5 Sing Sung Rd., Sec. 3, TAIPEI 107. ☎ 321.0698

TURKEY – TURQUIE
Librairie Hachette,
469 Istiklal Caddesi, Beyoglu, ISTANBUL. ☎ 44.94.70
et 14 E Ziya Gökalp Caddesi, ANKARA. ☎ 12.10.80

UNITED KINGDOM – ROYAUME-UNI
H.M. Stationery Office, P.O.B. 569,
LONDON SE1 9 NH. ☎ 01-928-6977, Ext.410
or
49 High Holborn, LONDON WC1V 6 HB (personal callers)
Branches at: EDINBURGH, BIRMINGHAM, BRISTOL,
MANCHESTER, CARDIFF, BELFAST

UNITED STATES OF AMERICA
OECD Publications Center, Suite 1207, 1750 Pennsylvania Ave.,
N.W. WASHINGTON, D.C.20006. ☎ (202)298-8755

VENEZUELA
Libreria del Este, Avda. F. Miranda 52, Edificio Galipán,
CARACAS 106. ☎ 32 23 01/33 26 04/33 24 73

YUGOSLAVIA – YOUGOSLAVIE
Jugoslovenska Knjiga, Terazije 27, P.O.B. 36, BEOGRAD.
☎ 621-992

Les commandes provenant de pays où l'OCDE n'a pas encore désigné de dépositaire peuvent être adressées à :
OCDE, Bureau des Publications, 2 rue André-Pascal, 75775 PARIS CEDEX 16.
Orders and inquiries from countries where sales agents have not yet been appointed may be sent to:
OECD, Publications Office, 2 rue André-Pascal, 75775 PARIS CEDEX 16.

OECD PUBLICATIONS
2, rue André-Pascal, 75775 Paris Cedex 16

No. 40.329 1978

PRINTED IN FRANCE